TEACH YOURSELF BOOKS

ELECTRICITY

This book provides a clear and concise introduction to the basic principles of electricity, explaining the meaning of electrical terms and showing how to tackle electrical calculations. It is equally suitable for those who are new to the subject and those who wish to revise half-forgotten knowledge. Special consideration is given to the needs of the private student who has few opportunities for practical work. SI units are used throughout.

TEACH YOURSELF BOOKS

ELECTRICITY

C. W. Wilman
M.Sc., Ph.D., C.Eng., F.I.E.E.

TEACH YOURSELF BOOKS
HODDER & STOUGHTON

ST. PAUL'S HOUSE WARWICK LANE
LONDON EC4P 4AH

First printed 1942
Second edition 1969
Third edition 1973
Third impression 1975

Copyright © 1969, 1973 editions
The English Universities Press Ltd

ISBN 0 340 05569 3

Printed in Great Britain
for Teach Yourself Books, Hodder & Stoughton,
by Richard Clay (The Chaucer Press), Ltd., Bungay, Suffolk

INTRODUCTION

THIS book has been written for the reader who seeks to learn something of electrical principles but who is dependent upon his own efforts to obtain and make use of the necessary information. It is hoped that it will be of use not only to those who have yet to take the first step towards an electrical career, but also to the practical electrician who has had little opportunity to gain a proper theoretical foundation for his work.

Attention has been paid also to the needs of the reader who is interested in the study of electricity for its own sake, or because he is engaged in a non-technical capacity in the electrical industry. Finally, it may be that *Teach Yourself Electricity* will prove of value to members of the Services, both for the immediate needs of those in the technical branches and for those who look forward to doing electrical work later on.

The subject is a wide one, and the requirements of different classes of reader are diverse. Moreover, it is important that the student who has not access to an electrical laboratory should approach his studies in a way which does not place him at too great a disadvantage in comparison with those more fortunately situated. These considerations have been carefully weighed in deciding the contents of our book and the order in which they should be presented.

A stumbling-block to many who take up a technical subject is the question of mathematics. Every endeavour has therefore been made to treat each aspect of electrical science in the manner that allows it to be followed with the least mathematical effort. For the most part, the only mathematics expected are a moderate knowledge of arithmetic and a slight acquaintance with elementary algebra. Wherever possible, results have been put into words rather than into mathematical formulæ.

The most difficult part of ordinary electrical work to the non-mathematical reader is the study of alternating current, but to omit all reference to this important subject would be to leave too large a gap in our survey. Those to whom the chapters on alternating current present special difficulty, and all who wish to pursue the matter further, will find much that is useful in the book in this series called *Teach Yourself Trigonometry*.

The student of electricity should take every opportunity to improve his mathematical equipment, and in particular should beware of the use of formulæ looked up in reference books and employed without any adequate understanding of the principles upon which they are founded. Special attention should be paid to the numerical examples in this book, both those worked in the text and those included in the questions at the end of each chapter.

Later on, the reader who is anxious to make further progress will need more specialized books, and will find plenty of them. We trust that by that time he will have decided that it is not an unprofitable occupation to *Teach Yourself Electricity*.

C. W. W.

NOTE TO 1973 EDITION

In the 1969 edition of this book, reference was made to the *Système International d'Unités* (SI). This system of units is now firmly established, and its use is continued and extended in the present edition. A list of the units included is given on page 179.

C. W. W.

CONTENTS

FIGURE TITLES

CHAPTER I

THE ELECTRIC CURRENT

It is convenient to begin our study of electricity by examining some of the properties of an electric current, and to leave to a later stage any enquiry as to what the current really is. We are thus enabled to deal at the outset with established facts and to proceed by easy stages from the familiar to the unfamiliar. The laws governing the practical applications of electric current are well established, but the last word on the nature of electricity itself is far from being said.

For our present purpose it is sufficient to look upon the electric current as something that flows. Just as water will flow along a pipe, so electricity will pass along a suitable path. Unlike the current of water, however, the electric current does not require a hollow pipe for its transmission; it is sufficient to provide a solid path of appropriate material.

Electric Conductors and Insulators

By "appropriate material" in the preceding paragraph we mean one that does not offer undue resistance to the flow of current. A pipe with a rough internal surface will not allow water to flow so readily as one in which the internal surface is smooth, and although there is no question of an internal surface in the case of an electric conductor, it remains true that different substances vary enormously in the resistance which they offer to the passage of an electric current.

The best conductor of electricity is silver. Copper is nearly as good and far less expensive; it is therefore the conductor most often employed in practical work. Most of the pure metals are moderately good conductors, and aluminium is a useful alternative to copper for overhead lines.

Sometimes, in order to limit the flow of current, a conductor is required which will offer a greater resistance than the ordinary pure metals, and for this purpose special alloys have been developed. Such materials may offer a resistance thirty or more times that of copper.

Other substances have higher resistances; that of carbon, for example, is about three thousand times that of copper. Finally, we reach a class of materials the resistance of which is so high that they are called non-conductors or insulators. The resistance of paraffin wax is many million million times that of copper. Other good insulators are suitable grades of ebonite (hard rubber), bakelite, porcelain, and dry silk and cotton. All these are extensively used for keeping apart conductors which might otherwise touch, and so form unwanted paths for the current.

In recent years much attention has been paid to materials known as semi-conductors, with resistances far greater than those of good conductors but far less than those of good insulators. A full consideration of their many unusual properties lies beyond the scope of this book, but simple examples occur on pages 26 and 152.

It should be noted that the terms " conductor " and " insulator " are only relative; all conductors offer *some* resistance and all insulators will pass *some* current. But the extremes are so far apart that there is every justification for looking upon materials at one end of the list as conductors and those at the other end as insulators.

Effects of Current

There are three main effects by which the presence of a current may be recognized: the heating effect, the magnetic effect and the chemical effect. All are of great importance in the practical applications of electricity. We might add a fourth: the physiological effect of the current which we term an electric shock.

Heating Effect of Current

The heating effect is a direct result of the resistance of the conductor through which the electricity flows. In order to force the current against the resistance, energy must be expended, and this energy appears in the conductor in the form of heat. In a similar manner, water forced along a pipe with a rough internal surface tends to heat the pipe, owing to the friction set up with its walls. Clearly, the heating effect of a given current will be greatest, other things being equal, when the substance through which it is made to flow is one that offers a high resistance.

Simple Magnetism

Between electricity and magnetism there are intimate associations which we shall have to study in some detail in a later chapter; for the present we need consider the subject only in its simplest aspect.

A magnet is a body having the property of setting up a magnetic field in the surrounding space. The existence of this field can be demonstrated by laying a

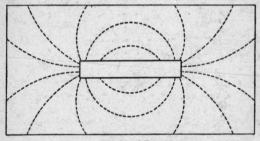

FIG. 1.—Field of Bar Magnet.

flat sheet of paper or thin card over the magnet and dusting fine particles of iron (such as iron filings) on to this surface. Many of the particles, as might be expected, take up positions over the edges and near the ends of the magnet, but others form a pattern extending some distance away.

The pattern obtained in the case of a simple bar magnet is shown in Fig. 1. If sufficient iron filings are used, there will be many more lines than are actually shown, but the general arrangement will be the same. It is clear that the magnetic field has a definite direction, just as the grain in a log of wood has a definite direction, although the direction may vary from point to point.

A comparatively light and free magnetized body, such as a compass needle, tends to set itself in the direction of any magnetic field in which it is situated. A compass needle placed near the magnet in Fig. 1 would set itself in the direction of the dotted lines at the position in which it was placed.

The reason why, in the absence of any disturbing influence, a compass needle points roughly north and south is that it sets itself in the direction of the magnetic field produced by the earth, which itself acts as a large magnet. In most parts of the world the direction of the field differs from true north and south, and we therefore speak of the direction in which the compass needle sets itself as magnetic north and south.

A compass needle can be deflected from its normal position by magnetic bodies, which tend to alter the direction of the earth's field in their vicinity. It is for this reason that special precautions have to be taken in mounting compasses on board ship.

Magnetic Effect of Current

A conductor carrying an electric current is surrounded by a magnetic field which interacts with the earth's

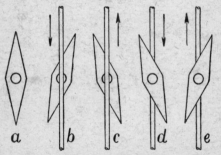

FIG. 2.—Magnetic Effect of Current.

field, thereby producing a change in the direction of the field in the vicinity of the conductor. This change can be detected by a compass needle.

In Fig. 2 (*a*), a compass needle is shown pointing to magnetic north and south in the earth's field. If a wire or other conductor is placed over the needle as shown at *b*, and the wire carries a current flowing in the direction of the arrow, the needle is deflected as shown. If the direction of the current is reversed, the direction of deflection is reversed also, as shown at *c*. If in either case the wire had been placed *under* the needle, the deflection would have been in the opposite direction, as shown at *d* and *e*.

A more pronounced effect can be obtained by combining the effect of *b* and *e*, or *c* and *d*. This can be done by doubling the wire back underneath the needle. The magnetic fields from both parts of the wire then assist each other in deflecting the needle in the same direction.

The direction of the magnetic field around a single con-

 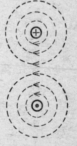

Fig. 3.—Field round Conductor.   Fig. 4.—Effect of two
Conductors.

ductor conveying a current is shown at Fig. 3. In this drawing, the small circle represents the conductor in section, and the current is supposed to be flowing down it into the paper. The cross on the wire, representing the rear view of an arrow, indicates this condition in a conventional manner.

Fig. 4 shows two parallel wires with current flowing in opposite directions. The dot on the lower wire represents the point of an arrow and indicates that the current is coming up out of the paper. It will be noted that the magnetic fields from the two wires assist each other in the space between them. This explains the increased effect on the compass needle when the wire is doubled back on itself with the needle between the two sections.

In these drawings the arrows on the dotted lines representing the field are merely conventional indications showing the direction in which the end of a compass needle which normally points to the north would point if the needle were influenced solely by the field in question.

FIG. 5.—Solenoid.

Consider the case of a length of wire wound into a coil, the turns lying side by side as shown in Fig. 5.

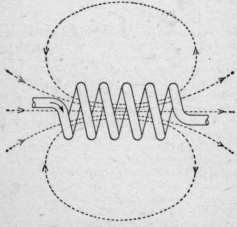

FIG. 6.—Field of Solenoid.

Such a coil is termed a **solenoid**. Suppose that current flows through the wire in the direction indicated by the arrow—*i.e.*, in the same direction as the movement of the hands of a clock when the coil is viewed from its left-hand end.

The direction of the magnetic field round the individual turns will then be as indicated by the dotted circles. Between each turn and the next the two fields are in opposition, and tend to cancel out. This is particularly true in the case of a coil having its turns close together, and the field of such a coil therefore assumes the form shown in Fig. 6.

It will be observed that this field is similar to that of the bar magnet in Fig. 1, and the coil carrying current is, in many respects, equivalent to a bar magnet. It will have a similar effect, for example, upon a compass needle. Moreover, if an unmagnetized bar of iron or steel be inserted in the coil, it will become magnetized, and will remain so while the current continues to flow. Whether or not it will remain a magnet after the flow of current ceases depends upon the material of which it is made; we shall have to return to this question in Chapter VII.

Chemical Effect of Current

Liquids, like solids, vary enormously in the resistance which they offer to the passage of current. Many oils, for example, are exceptionally good insulators, while at the other end of the scale mercury, like most metals, is a reasonably good conductor. Water, if pure, is a very poor conductor, but the least trace of impurity sends down its resistance, so that in the form in which it is normally available it is quite possible to pass current through it.

When a current flows through a solid conductor, there is usually no change in the chemical nature of the conductor; if it is copper, for example, it goes on being copper indefinitely. Some liquids, however, are affected chemically by the passage of a current. If a little sulphuric acid is added to water in order to improve its conductivity, the water can be split up into its constituents, oxygen and hydrogen, by the passage of a current. The oxygen collects at the point at which the current enters the liquid, and the hydrogen at the point at which it leaves it.

Many solutions of chemical salts in water undergo similar changes. These liquids are known as **electrolytes**, a word which means decomposable by electricity, and the process is known as **electrolysis**.

If the current is conducted into and out of the liquid by means of immersed metal plates, the substance of these plates may also be acted upon, metal being removed from one plate and deposited upon the other. This is the basis of electroplating.

Production of Electric Currents

We have seen that the passage of an electric current can give rise to heat, to magnetic action, and to chemical changes. In due course we shall find that these effects are reversible: just as a current can give rise to heat, so can heat give rise to a current; just as a current can produce magnetic action, so can magnetic action produce a current; and just as a current can bring about chemical changes, so can chemical changes bring about a current.

Detection and Measurement of Current

Each of the three main effects of a current can be employed under suitable conditions to detect its presence or to measure its strength. One kind of measuring instrument depends upon the heating of a conductor through which the current flows, while the strength of a current can be still more accurately determined by the amount of chemical action which it will perform in a given time.

The magnetic effect, however, is the most important for purposes of detection and measurement. We have seen that a compass needle can be deflected by current in an adjacent conductor, and this fact is made use of in one form of the detecting and measuring instrument known as a **galvanometer**.

A simple galvanometer is shown in Fig. 7. In place of a single loop of wire we now have a number of turns wound as a hollow coil (*a*) surrounding the compass needle (*b*), the latter being provided with a pointer (*c*) in order that its movements may be readily followed. The parts of the turns lying above the needle and the parts lying below the needle tend, as we have already seen, to deflect it in the same direction, and as a large number of turns can be used, the instrument can be made fairly sensitive.

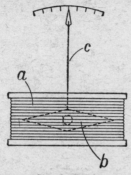

FIG. 7.—Simple Galvanometer.

As in Fig. 2 on page 14, the needle makes a clockwise movement (like the hands of a clock) in response to current flowing in one direction, and a counter-clockwise movement (opposite to the hands of a clock) in response to current flowing in the other. It is therefore possible to detect not only the presence of a current, but also the direction in which it is flowing.

Since the galvanometer shown in Fig. 7 is dependent upon the magnetic field of the earth, it must be set so that the needle lies north and south when no current is flowing. It is therefore liable to disturbance by external magnetic fields, and for these reasons more elaborate instruments, to be described later, are usually employed.

Unit of Current

We have spoken of the measurement of current, but obviously we cannot do any measuring unless we have some standard of current strength. Just as we need an agreed unit of length in order to measure the length of a pole, so we need an agreed unit of strength in order to measure the strength of a current. It is not very important what unit is chosen, so long as everybody uses the same one.

The unit of current strength which has been universally agreed upon is the **ampere**.* It takes its name from André Marie Ampère, a French scientist who studied magnetism and electricity over a century ago.

It is important to realize exactly what it is that we measure in amperes (commonly abbreviated amps or A). Consider the case of a current of water. Clearly, it would be meaningless to speak of a current of so many cubic metres, because we should be left in ignorance of the rate at which the cubic metres were coming past. But if we speak of a current of so many cubic metres per hour, we have a clear idea of the size of the stream with which we have to deal. The ampere is a convenient way of expressing in the case of electricity the idea conveyed by the term " cubic metres per hour " in the case of water.

The electrical unit of quantity is the **coulomb** † (named after Charles Augustin de Coulomb, another early experimenter), and the strength of a current in amperes is equal to the number of coulombs passing per second.

The following examples will give some idea of the values of current in amperes that we are likely to meet:

> An average domestic heater takes about six amperes.
>
> An ordinary electric lamp takes under half an ampere.
>
> A street cable may carry several hundred amperes.
>
> A very simple galvanometer will detect a current of one thousandth of an ampere.
>
> More elaborate galvanometers will detect a current of considerably less than one ten-millionth of an ampere.

For the measurement of small currents it is convenient to have a unit smaller than the ampere, and the **milli-ampere** (abbreviated milliamp or mA) is used. The milliampere is equal to one thousandth of an ampere.

QUESTIONS

1. What is meant by a good conductor of electricity ? Name one.

* Pronounced *am-pair*.　　　　　† Pronounced *koo-lom*.

2. What is meant by a good insulator? Name one.

3. What are the principal effects of an electric current?

4. How can the magnetic effect of a current be demonstrated?

5. What happens when a current flows through an electrolyte?

6. How can the presence of a current be detected?

7. How can the direction in which a current flows be determined?

8. What is the unit of current, and what, exactly, does it measure?

9. How many milliamperes are there in 0·016 amp?

10. How many amperes are there in 2675 milliamps?

11. How many coulombs per second does a current of 500 milliamps represent?

CHAPTER II

ELECTRICAL RESISTANCE

THIS chapter and the next are devoted largely to simple electrical calculations. As some familiarity with this part of the subject is essential to further progress, the reader is urged to work carefully through each example.

Unit of Resistance

We have seen that different conductors offer different resistances to the flow of current, but before we can compare one conductor with another in this respect, we must have a unit in which resistance can be measured.

The agreed unit is the ohm, named after Georg Simon Ohm, a German scientist. The Greek capital letter omega (Ω) is commonly used as an abbreviation for this quantity, so that the expression " 10Ω " means a resistance of 10 ohms.

The range of resistances to be measured is so extensive that it has been found convenient to employ two further units. The megohm, equal to a million ohms, is used for the measurement of very high resistances, and the microhm, equal to one millionth of an ohm, for the measurement of very small ones. We have, therefore,

1,000,000 microhms = 1 ohm
1,000,000 ohms = 1 megohm.

The following are some examples of resistance values met with in practice:

A copper wire one millimetre in diameter and 50 metres long has a resistance of about 1 ohm

An ordinary household electric lamp has a working resistance of up to 4000 ohms.

The resistance offered by the insulation between one conductor and another is seldom less than a megohm and is usually much more.

Size of Conductor

A large number of people trying to pass along a narrow corridor find it difficult to get through. They would obviously have less difficulty if the width of the corridor were increased.

In an analogous manner, the resistance offered by a conductor to an electric current is dependent upon its cross-sectional area. The greater the cross-sectional area, the less the resistance, and vice versa.

The term "cross-sectional area" means the area which would be exposed by cutting straight across the conductor; in the case of a circular wire, for example, it is the area of a circle of the same diameter as the wire.

FIG. 8.—Cross-Sectional Area.

In mathematical language, we may say that the resistance is inversely proportional to the cross-sectional area. By this we mean that if the cross-sectional area is halved, the resistance is doubled, and so on.

Note that the area of a circle is proportional to the square of its diameter (*i.e.*, to the diameter multiplied by itself), so that if the diameter of a circular wire is halved, the cross-sectional area is divided by four, and so on. Thus, in Fig. 8, each of the small circles is one quarter the area of the large circle.

Example.—*A certain wire has a diameter of one millimetre and a resistance of half an ohm. What will be the resistance of a wire having a diameter of one-third of a millimetre but otherwise similar to the first one?*

Diameter of second wire is $\frac{1}{3}$ that of the first.
Cross-sectional area is $\frac{1}{3} \times \frac{1}{3} = \frac{1}{9}$ that of the first.
Resistance of second wire $= 9$ times the resistance of the first $= (9 \times 0.5)$ ohms $= 4.5$ ohms.

Length of Conductor

If a length of, say, 1 metre of a uniform conductor offers a certain resistance to the flow of current, the next metre of the conductor will naturally offer the same resistance. It follows that if other things are equal the resistance of a conductor is proportional to its length.

Example.—*A wire 2 kilometres long has a resistance of 88Ω. What is the resistance of a piece of this wire one metre long?*

2000 metres offer a resistance of 88Ω.
1 metre offers a resistance of 0.044Ω.

Resistivity

We have seen in Chapter I that the material of which a conductor is made is an important factor in determining its resistance. In order to compare different materials in this respect, it is usual to state the resistance offered by a cube having a side of 1 metre, it being assumed that current is passed through the cube from one face to the opposite one. This figure is known as the resistivity of the material in question.

The following are some examples of resistivity:

Silver	.	.	0.016 microhms per metre cube
Copper	.	.	0.017 ,, ,, ,, ,,
Iron	.	.	0.12 ,, ,, ,, ,,
Carbon	.	. 50	,, ,, ,, ,,

Quite a small amount of impurity in a metal will cause a considerable increase in its resistivity. Alloys of two or more metals usually have a higher resistance than would be expected from the values of their constituents.

Care must be taken to distinguish between the terms "metre cube" and "cubic metre". The metre cube certainly has a volume of 1 cubic metre, but it also has a definite shape—*i.e.*, a length of 1 metre and a cross-sectional area of 1 square metre. The cubic metre is a measure of volume only, and this by itself is no guide to resistance.

If we know the resistivity of a material, it is an easy matter to calculate the resistance of any conductor made of that material. The resistivity is the resistance of a conductor 1 metre long and 1 square metre in cross-sectional area, and we have seen that resistance is proportional to length and inversely proportional to cross-sectional area. We therefore multiply the resistivity by the length of the conductor in metres and divide the result by the cross-sectional area in square metres.

Example.—*What is the resistance of a pure copper wire 50 metres long and 1 square millimetre in cross-sectional area?*

Length = 50 *metres.*
Cross-sectional area = 1 *square millimetre* = 0·000001 *square metre.*
Resistivity of copper = 0·017 *microhm,* = 0·000000017 *ohm, per metre cube.*
Resistance of wire

$$= \frac{resistivity \times length\ in\ metres}{cross\text{-}sectional\ area\ in\ square\ metres}$$

$$= \frac{0\cdot000000017 \times 50}{0\cdot000001} = 0\cdot85\ ohm.$$

The Greek letter ρ (rho) is often used as a symbol for resistivity. If we denote the length of a conductor by l and the cross-sectional area by A, we may therefore abbreviate the above statement of the resistance of any conductor by writing:

$$\text{Resistance} = \rho\frac{l}{A}.$$

Resistivities are sometimes quoted for a centimetre cube instead of for a metre cube, and if such figures are used, the length of the conductor must be expressed in centimetres and the cross-sectional area in square centimetres before proceeding with the calculation.

Conductance

A good conductor offers a low resistance, and a poor conductor offers a high resistance. The term **conductance** is therefore used as the opposite of resistance, and we may write:

$$\text{Conductance} = \frac{1}{\text{Resistance}},$$

from which it follows that as resistance goes down, conductance goes up.

The idea of conductance is used mainly in rather advanced calculations, but later on in the chapter we shall find that even in simple problems it is sometimes more convenient to think in terms of conductance than in terms of resistance. All that we need to remember at present is that, as expressed in the above equation, conductance is the reciprocal of resistance.

Effect of Temperature

The resistance of nearly all substances varies appreciably as the temperature changes. In the case of most conductors, a *rise* in temperature results in an *increase* in resistance.

Tables of resistivities usually assume a temperature of 0° Celsius (*i.e.*, Centigrade), and for accurate calculations it is necessary to make a correction according to the actual working temperature.

For moderate variations it is sufficient to add a definite proportion of the resistance at 0° C for each degree rise in temperature. This proportion is termed the **temperature coefficient** of resistance.

Example.—*The temperature coefficient for copper is 0·004 per degree C, and the resistance of a certain copper wire at 0° C is 100 Ω. What will its resistance be at 20° C?*

Resistance at 0° C = 100 Ω.
Temperature rise = 20° − 0° = 20°.
Increase in resistance = (100 × 0·004 × 20) Ω = 8 Ω.
Resistance at 20° C = 108 Ω.

The resistance of some substances, including carbon,

electrolytes and insulators, becomes smaller instead of greater as the temperature rises. Such substances are said to have a negative temperature coefficient, and in their case the variation in resistance for a rise in temperature must be subtracted instead of added.

Materials of comparatively high resistivity (semi-conductors) having large negative temperature coefficients have been developed for special purposes. The resistance of such materials can be made very sensitive to temperature changes.

Certain alloys have an almost negligible temperature coefficient, and are therefore of value in the construction of standard resistances with which unknown resistances are to be compared.

The following are some typical temperature co-efficients:—

Silver	. .	0·004 per degree C
Copper	. .	0·004 ,, ,,
Iron	. .	0·006 ,, ,,

The figure for the alloys referred to in the preceding paragraph may be as low as 0·00002 per degree C, and for practical purposes the resistance of these alloys can be regarded as unaffected by temperature changes.

Since resistance is a quantity which can be accurately measured, the variation in resistance of a conductor affords a convenient means of measuring temperatures too high to be dealt with by an ordinary thermometer. Instruments embodying this principle are termed resistance pyrometers.

By the use of very elaborate cooling arrangements, temperatures can be reduced to a point at which the resistance of some conductors almost disappears. This effect, which is of considerable theoretical interest, is known as super-conductivity.

Series and Parallel Connexions

There are two simple ways in which pieces of electrical apparatus can be connected. If they are arranged so that the current passes first through one and then through the next, they are said to be connected in series. If they

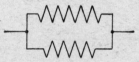

are arranged so as to offer alternative parallel paths to the current, they are said to be connected **in parallel**.

FIG. 9.—Resistance Symbol.

The symbol shown in Fig. 9 is used to denote a resistance. Two resistances connected in series are shown in Fig. 10, and two resistances connected in parallel in Fig. 11.

FIG. 10.—Resistances in Series.

FIG. 11.—Resistances in Parallel.

Resistances in Series

Resistances in series can be compared to the separate portions of a single conductor. The total resistance is therefore the sum of the individual resistances.

Example.—*What is the total resistance of resistances of 45, 5 and 9 ohms connected in series ?*

Total resistance = (45 + 5 + 9) *ohms* = 59 *ohms.*

Resistances in Parallel

As suggested in Fig. 12, a single conductor can be looked upon as two conductors, each of one half the

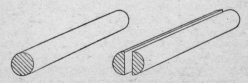

FIG. 12.—Parallel Conducting Paths.

total cross-sectional area, joined in parallel. Since resistance is inversely proportional to cross-sectional area, we conclude that the joint value of two equal resistances in parallel is one half that of one of them, and similarly for other numbers.

Example.—*What is the joint resistance of 4 resistances, each of 20 ohms, connected in parallel?*

$$Joint\ resistance = \frac{20}{4}\ ohms = 5\ ohms.$$

This method is obviously inapplicable if the values of the resistances are unequal, but by noting that the parallel connexion results in an increase in the conductance, we are led to a more general solution. All that is necessary is to find the reciprocal of each resistance value, add the reciprocals, and then find the reciprocal of the result.

The modern unit of conductance is the **siemens*** (singular and plural). The number of siemens is simply the reciprocal of the number of ohms. The name is that of several eminent electrical engineers of the nineteenth century.

Example.—*What is the joint value of resistances of 3 ohms and 6 ohms connected in parallel?*

> *Resistances = 3 ohms and 6 ohms.*
> *Conductances = 0·333 and 0·167 siemens.*
> *Sum of conductances = 0·5 siemens.*
> *Joint resistance = 2 ohms.*

Example.—*What resistance must be placed in parallel with one of 2 ohms to produce a joint resistance of 1·5 ohms?*

> *Joint resistance = 1·5 ohms.*
> *Conductance = 0·667 siemens.*
> *Existing resistance = 2 ohms.*
> *Conductance = 0·5 siemens.*

Difference of conductances = 0·167 siemens.
Extra resistance required = 6 ohms.

* Pronounced *seem-ens*.

Those who prefer to work to a formula will find from these examples that the parallel value of, say, three resistances, a, b and c, is

$$\frac{1}{\frac{1}{a} + \frac{1}{b} + \frac{1}{c}}$$

and similarly for any other number. In the case of two resistances, the expression can be reduced to the simpler form

$$\frac{ab}{a + b},$$

that is, the product divided by the sum. As an exercise, the above calculations may be checked by each of these formulae.

Note that the joint resistance of two or more resistances in parallel is always less than the least single resistance, no matter how large the others may be.

Note also that the value of two equal resistances in series is four times that of the same resistances in parallel.

Graphical Solution for Parallel Resistances

FIG. 13.—Joint Parallel Resistance.

There is a simple graphical method of finding the joint value of resistances in parallel. It is illustrated in Fig. 13 for the first of the examples given above.

Draw a horizontal base-line of any convenient length, and then erect a vertical line at each end, making the length of each vertical line proportional to one of the resistances. Thus, in our case, one vertical line is six units long and the other three units.

Join the top of each vertical line to the base of the other, and from the point where these sloping lines cross, draw a third vertical line to the base. The length of this third vertical line represents the joint resistance. In our case its length is two units, which agrees with the result already obtained by calculation.

If there are more than two resistances, the joint resistance of any two may be found first and the result combined with the next, and so on.

Series and Parallel Groups

Combinations of series and parallel resistances can often be evaluated by dividing them into groups. Thus,

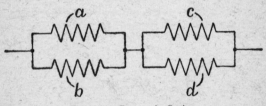

FIG. 14.—Groups in Series.

in Fig. 14 we consider a and b as one group, and c and d as another. The total resistance is therefore the parallel

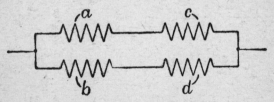

FIG. 15.—Groups in Parallel.

resistance of a and b added to the parallel resistance of c and d.

In Fig. 15, on the other hand, we add a to c and b to d, and then find the parallel resistance of the results.

Resistors

Any device (for example, a length of wire) the primary object of which is to offer resistance is termed a **resistor**.

QUESTIONS

1. In what units is electrical resistance measured?

2. How do the dimensions of a conductor affect its resistance?

3. How does the material of which a conductor is made affect its resistance?

4. How does the temperature of a conductor affect its resistance?

5. What do you understand by (a) resistivity and (b) temperature coefficient of resistance?

6. What is meant by conductance?

7. What is the joint value of resistances of 90 ohms and 10 ohms connected in parallel?

8. What is the total resistance of resistances of 90 ohms and 10 ohms connected in series?

9. A resistance of 200 ohms is required, and resistances of 50 ohms and 60 ohms are available. What third resistance must be used in order that the three may have the required value when connected in series?

10. What is the joint value of resistances of 0·4 ohm and 0·6 ohm connected (a) in series, (b) in parallel?

11. What is the joint resistance of resistances of 2, 5 and 10 ohms connected in parallel?

12. To an existing resistance of 20 ohms, a resistance of 80 ohms is connected in parallel. What resistance must be connected in series with the combination in order to restore the original value?

13. What resistance must be placed in parallel with one of 1·5 megohms to produce a joint resistance of one megohm?

14. The resistance of a certain copper conductor is 25 ohms at a temperature of 0° C. What is its resistance at 20° C? (Assume that temperature coefficient of copper is 0·004 per °C.)

15. Four resistors have the following values:

(a) 2 ohms.
(b) 6 ohms.
(c) 4 ohms.
(d) 8 ohms.

Find their joint resistance when connected (i) as shown in Fig. 14, (ii) as shown in Fig. 15.

CHAPTER III

VOLTAGE AND OHM'S LAW

Electromotive Force

In order to cause a current to flow against a resistance, a driving force is necessary. This force is called **electromotive force** or **e.m.f.** Its symbol is E.

At a later stage we shall find that electromotive force can be produced by various devices for converting chemical, mechanical or heat energy into electrical energy.

Potential Difference

When, as the result of the application of electromotive force, current flows through a resistance, a difference of potential is said to exist between the two ends of the resistance. **Potential difference** is denoted by the abbreviation **p.d.** Its symbol is V.

If two water-tanks are connected by a pipe, a current of water will flow through the pipe so long as the level of one tank is above that of the other. The difference in level corresponds to the difference of potential in the electrical case. If there is no difference of level between the two tanks, or no difference of potential between the two ends of an electrical conductor, there will be no current.

Unit of E.M.F. and P.D.

Although it is convenient to make a distinction between electromotive force and potential difference, these quantities are of a similar nature and can be measured by the same unit.

This unit is the **volt** (named after Alessandro Volta, an Italian physicist), and both electromotive force and potential difference are often referred to as **voltage.**

The **millivolt,** equal to one thousandth of a volt, is used for the measurement of very small potential differences, and the **kilovolt,** equal to 1000 volts, for the measurement of very large ones. We have, therefore,

$$1000 \text{ millivolts} = 1 \text{ volt}$$
$$1000 \text{ volts} = 1 \text{ kilovolt.}$$

The following are examples of voltages commonly encountered:—

An electric torch battery has an e.m.f. of from $1\frac{1}{2}$ to $4\frac{1}{2}$ volts.

Most domestic electricity supplies are of 240 volts.

The voltage in a wireless receiving aerial may be only a few millivolts.

Overhead power-transmission lines may be designed for 132 kilovolts or more.

Ohm's Law

Consider the resistance R shown in Fig. 16, and suppose that a current is flowing through it in the direction of the arrow. The letter I is employed as a symbol for current.

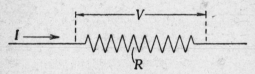

FIG. 16.—Resistance carrying Current.

Since a current is flowing through a resistance, there must be a difference of potential. This difference of potential may be looked upon as a voltage existing *across* the resistance, as indicated by V.

If the voltage is increased, it is natural to expect that more current will be forced through the resistance, and this is actually the case. If, on the other hand, the resistance is increased, the current will be reduced.

It is found that the current is proportional to the voltage and inversely proportional to the resistance. In other words, doubling the voltage doubles the current, while doubling the resistance halves the current, and so on.

Provided that the units in which the quantities are measured are properly chosen, we may express this relationship by writing:

$$\text{Current} = \frac{\text{Voltage}}{\text{Resistance}},$$

or, in symbols,

$$I = \frac{V}{R}.$$

This important result is known as Ohm's Law, and it enables us to calculate any one of the three quantities when the other two are known.

The units ampere, ohm and volt are such that if a current of 1 ampere flows through a resistance of 1 ohm the potential difference is 1 volt. We may therefore memorize Ohm's Law in the form :

$$\text{Amperes} = \frac{\text{Volts}}{\text{Ohms}}.$$

Example.—*What current is flowing through a resistance of* 100 *ohms the potential difference across which is* 20 *volts ?*

$$I \ (current) = \frac{V \ (voltage)}{R \ (resistance)} = \frac{20}{100} \ amp. = 0.2 \ amp.$$

The equation:

$$\text{Current} = \frac{\text{Voltage}}{\text{Resistance}}$$

can be written in two other forms, thus:

$$\text{Voltage} = \text{Current} \times \text{Resistance}$$

and $$\text{Resistance} = \frac{\text{Voltage}}{\text{Current}}.$$

From one or other of these we can find either the current, the voltage or the resistance. As an aid to memory we may write:

$$\frac{V}{IR}.$$

If any one of these letters is covered up, the other two give the value of the quantity in question. Thus:

$$V = IR,$$

$$I = \frac{V}{R},$$

and $$R = \frac{V}{I}.$$

These expressions correspond to those already given in words.

Example.—*What potential difference exists across a resistance of 500 ohms through which a current of half an ampere is flowing?*

$$V = IR = (0.5 \times 500) \; volts = 250 \; volts.$$

Example.—*The potential difference across a resistance through which a current of 20 amperes is flowing is 35 volts. What is the resistance?*

$$R = \frac{V}{I} = \frac{35}{20} \; ohms = 1.75 \; ohms.$$

Graphical Representation of Ohm's Law

If for a given resistance we work out the currents corresponding to different potential differences and plot the results on squared paper, we obtain a number of

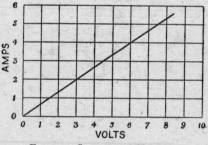

FIG. 17.—Current and Voltage.

points lying on a straight line. Fig. 17 shows the result for a resistance of 1.5 ohms, the figures along the base of the rectangle representing volts and those on the left-hand side amperes.

From such a diagram, the voltage corresponding to any given current can be read off directly.

The fact that the points lie on a straight line is an indication that the current is proportional to the voltage. We shall have occasion to return to this result at a later stage.

Voltage across Series Resistances

Ohm's Law is true for any part of a circuit, and it can therefore be applied to any resistance of a group connected either in series or parallel.

If resistances are connected in series (Fig. 10), the current must be the same in all of them, and the voltage across each one is therefore proportional to its resistance value. The potential difference across a resistance forming part of a series circuit is often spoken of as the **fall of potential** or **voltage drop** in that resistance.

Example.—*Resistances of 10 ohms and 120 ohms are connected in series. The voltage drop in the 10-ohm resistance is 15. What is the potential difference across the 120-ohm resistance ?*

P.D. across 10-ohm resistance = 15 *volts.*

$$,, \quad ,, \quad 120\text{-}ohm \quad ,, \quad = \left(15 \times \frac{120}{10}\right) volts$$

$$= 180 \ volts.$$

Current in Parallel Resistances

If resistances are connected in parallel (Fig. 11), the voltage is the same for them all, and the current in each is therefore inversely proportional to its resistance value.

Example.—*Resistances of 350 ohms and 150 ohms are connected in parallel. The current in the 350-ohm resistance is 0·3 ampere. What is the current in the 150-ohm resistance ?*

Current in 350-ohm resistance = 0·3 *amp.*

$$,, \quad ,, \quad 150\text{-}ohm \quad ,, \quad = \left(0\cdot3 \times \frac{350}{150}\right) amp.$$

$$= 0\cdot7 \ amp.$$

Note that if the voltage is maintained constant, the current in a resistance is not affected by the connexion of other resistances in parallel. The parallel arrangement is therefore that usually employed for the connexion of current-consuming devices to supply mains.

Consider, however, the case shown in Fig. 18. We will give the important points in the calculation and leave as an exercise the working out of the details.

Resistances of 24 and 30 ohms are connected in series. Neglect the third resistance for the moment, and suppose that the total voltage $V_1 + V_2$ is 108. V_1 is then 48 volts, V_2 is 60 volts and the current is 2 amperes.

Now let the 20-ohm resistance be added, as indicated by the dotted lines. The joint parallel resistance of the 30-ohm and 20-ohm resistances is 12 ohms, making with the 24-ohm resistance a total of 36 ohms.

The current is now 3 amperes, of which only 1·2 amperes flows through the original 30-ohm resistance. The voltage V_2 across this resistance is 36.

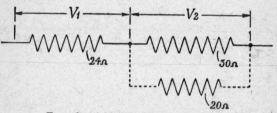

FIG. 18.—Added Parallel Resistance.

The addition of the parallel resistance has therefore altered the balance of the whole circuit, both the voltage across the original 30-ohm resistance and the current flowing through it being reduced. The reason for this is the presence of the 24-ohm series resistance, the voltage drop across which is increased owing to the extra current which flows when the 20-ohm resistance is added.

We conclude, therefore, that great caution must be exercised in the addition of parallel resistances to *part* of a circuit, since although the total voltage may remain constant, the way in which it is distributed over the different sections is likely to be changed.

Potential Divider

It is clear from the foregoing examples of series resistances that if we have a voltage causing a current to flow in a conductor, it is possible to obtain any smaller voltage by connecting to appropriate points on the conductor. Thus, in Fig. 19, a voltage V is applied to a conductor consisting of two resistances in series. By connecting leads a and b to the points shown, we obtain a smaller voltage V' proportional to the part of the resistance included between the tapping points.

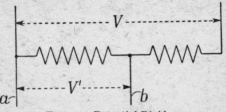

FIG. 19.—Potential Divider.

It is important to note, however, that the proportionality between the voltage obtained and the resistance between the tapping points is maintained only while the current drawn over the leads a and b is small enough to be neglected in comparison with that flowing over the main path. In other cases the conditions of Fig. 18 arise, and the voltage is reduced accordingly.

QUESTIONS

1. What is the distinction between electromotive force and potential difference?
2. In what units are e.m.f. and p.d. measured?
3. What are the numerical relations between a resistance, the voltage across it, and the current flowing through it?
4. Express Ohm's Law in a convenient form for calculating (a) current, (b) voltage and (c) resistance.
5. The voltage drop in a resistance of 25 ohms is 0·5. What is the current?
6. What voltage will cause a current of 1 milliampere to flow through a resistance of 1 megohm?
7. The potential difference across a resistance of 45 ohms is 15 volts. What is the current?
8. The current flowing through a resistance of 40 ohms is 2 amperes. What is the voltage across the resistance?

9. A potential difference of 1 volt will cause a current of 10 amperes to flow through a certain conductor. What is the resistance of the conductor?

10. Resistances of 400 ohms and 100 ohms are connected in parallel. If the current in the 400-ohm resistance is 1 ampere, what is the current in the other?

11. The currents in two parallel resistances are 4 amperes and 2 amperes respectively. What is the value of each resistance if the joint resistance is 50 ohms?

12. In Fig. 20, the voltage across the 36-ohm resistance

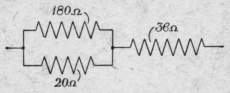

FIG. 20.—Series and Parallel Resistances.

is 90. Find (i) the current in the 180-ohm resistance, (ii) the current in the 20-ohm resistance, (iii) the voltage across the whole circuit.

CHAPTER IV

CHEMICAL EFFECTS OF CURRENT

IT was mentioned in Chapter I that some liquids, known as electrolytes, were decomposed by the passage of an electric current. We will now study this effect in rather greater detail.

Elements, Compounds and Solutions

All substances, liquids, gases and solids, are composed of minute particles, known as atoms. The atoms are usually collected into groups called molecules. There are less than a hundred known kinds of atoms, but the number of *combinations* of atoms, or molecules, is much greater.

A substance the molecules of which consist of only one kind of atom is known as an element. Copper, zinc, sulphur, mercury, oxygen, hydrogen, nitrogen and chlorine are examples of elements. Each element corresponds to a particular kind of atom.

A substance the molecules of which consist of more than one kind of atom is known as a chemical compound. A molecule of water, for example, consists of two atoms of the element hydrogen and one of the element oxygen. Water is therefore a chemical compound.

When two or more different atoms unite in this way, the properties of the original elements disappear and are replaced by new properties belonging to the chemical compound which they form. Thus, at normal temperatures, both hydrogen and oxygen are gases, but when atoms of hydrogen and oxygen unite to form water, the result is a liquid.

A chemical compound is therefore a more intimate association of elements than a mere mixture. It would be easy to form a mixture of hydrogen and oxygen by putting the two gases into the same container, but the result would not be water. Only when individual atoms are associated to form a new kind of molecule is a chemical compound formed.

A solution is an intimate mixture, but it is not a chemical compound. It is probable that when a chemical compound is dissolved, some at least of its molecules split up into separate atoms or groups of atoms. Of this nature are electrolytes.

Electrolysis

Fig. 21 shows two metal plates (or **electrodes**) immersed in an electrolyte. We will suppose that an electric current is passed through the device in the direction of the arrows.

The electrode *a*, at which the current enters the electrolyte, is termed the **anode**, and the electrode *b*, at which it leaves, the **cathode**.

By way of example, let both the anode and the cathode be formed of copper, and let the electrolyte be a solution of copper sulphate in water. Copper sulphate is a chemical compound of the elements copper, sulphur and oxygen.

Immediately the current commences to flow, copper from the dissolved copper sulphate appears at the cathode and is deposited there. This copper is replaced in the solution by an equivalent amount removed from the anode.

The result is that the cathode gains in weight while

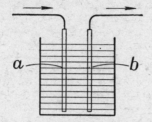

FIG. 21.—Electrolysis.

the anode loses. The gain in weight at the cathode is proportional to the current and the time, in other words, to the quantity of electricity which passes.

Applications of Electrolysis

A similar action takes place if, instead of the copper cathode, we employ some other conducting object. Thus, in Fig. 22, current is passed from the copper

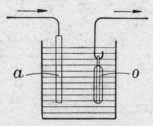

FIG. 22.—Electro-Plating.

anode *a* through the solution of copper sulphate to the suspended object *o*, and the latter, acting as the cathode, becomes copper plated.

Similarly, by the use of a silver anode and an electrolyte containing silver (such as a solution of silver nitrate) we can cause an article to be silver plated. Nickel, gold and chromium plating depend upon the same principle. In commercial plating work it is often found desirable to deposit first one metal and then another in order to obtain a coherent and durable coating.

In the printing industry the plating principle is used in the production of the printing blocks known as electrotypes.

A process very similar to electroplating is used on a large scale for refining copper. The ingots to be refined form the anodes, and thin sheets of copper the cathodes. When the current passes, pure copper is deposited on the cathodes, while impurities in the anodes either pass into solution or collect as sludge.

Electrolysis finds another practical application in the extraction of aluminium from the minerals in which it occurs. In this case, the raw material itself, in a molten state, forms the electrolyte. The aluminium is deposited at a carbon cathode (actually the lining of the furnace in which the action takes place) when a heavy current is passed through the molten material.

The fact that the amount of chemical action is proportional to the quantity of electricity passing (current × time) has been made use of in electricity meters. In one form, mercury is deposited from an electrolyte on to a cathode, from which it falls into a graduated tube. The level of the mercury in the tube indicates the quantity of electricity which has passed.

Chemical action can be employed to find the direction in which a current is flowing. We referred in Chapter I to the electrolytic decomposition of water into oxygen and hydrogen, the oxygen collecting at the point at which the current enters the liquid (anode), and the hydrogen at the point at which it leaves it (cathode). Since the volume of hydrogen is greater than that of oxygen, the anode and cathode can be recognized by the amount of gas given off.

The same object can be accomplished more readily by the use of **pole-finding paper**. This is paper soaked in a chemical solution which forms an electrolyte when the paper is moistened. If two wires between which there is a difference of potential are pressed on to the moistened paper a short distance apart, the chemical action is such that a mark is left against one of them, usually that at which the current enters.

Chemical Production of E.M.F.

Let us return to Fig. 21 and suppose that one electrode is now of zinc, the other of copper, and the electrolyte a dilute solution of sulphuric acid. Sulphuric acid is a chemical compound of the elements hydrogen, sulphur and oxygen.

Without any connexion to an external source of current, it is found that a voltage appears across the wires leading to the electrodes. We conclude that dissimilar metals immersed in an electrolyte form a means of generating an e.m.f. This arrangement is known as an **electric** or **voltaic cell**.

If the wires from the electrodes are joined outside the cell, the e.m.f. causes a current to flow through them from the copper to the zinc electrode. We might detect this current by means of a compass needle, as explained in Chapter I.

While the cell is supplying current, it is transforming chemical into electrical energy. In the process, hydrogen from the electrolyte collects at the copper electrode, and is replaced by zinc removed from the zinc electrode. The latter is therefore consumed, and the zinc may be looked upon as the fuel which enables the cell to do its work.

The connexion to the copper electrode (by which the current leaves the cell) is termed the **positive pole** of the cell, and the connexion to the zinc electrode (by which the current returns to the cell) the **negative pole**. In the external circuit, therefore, the current flows from the positive to the negative pole.*

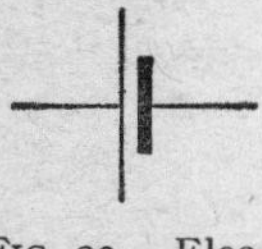

The marks + and − are often used to distinguish the two poles.

FIG. 23.—Electric Cell.

The symbol shown in Fig. 23 is used for a cell in circuit diagrams. The long, thin line represents the positive pole, and the short, thick one the negative.

* See also pages 174–5.

The e.m.f. of a cell is dependent upon the materials of the electrodes and the electrolyte and, to some extent, upon the temperature. It is not affected by the size of the cell. The combination we have described produces an e.m.f. of about 1 volt.

The following is a list of some materials which might be used as electrodes:

(1) Aluminium   (6) Copper.
(2) Zinc.   (7) Silver.
(3) Iron.   (8) Gold.
(4) Nickel.   (9) Carbon.
(5) Lead.

If any two of these materials are used to form a cell, the one lower in the list will, in general, be that at which the current leaves the cell, and the one higher in the list that which is consumed in the process of generating the current. The farther apart the two materials are in the list, the higher will be the voltage.

Primary and Secondary Cells

In the cell described in the preceding section, the zinc is consumed in the process of generating the current, and when it is used up it must be replaced by a new electrode. A cell of this kind is termed a **primary** cell.

With some combinations of electrodes and electrolyte, the electrodes merely suffer a chemical change during the production of current, and can be restored to their original condition by passing current from an external source through the cell in the reverse direction, *i.e.*, by electrolysis. Cells of this kind are termed **secondary** cells, and the process of restoring the electrodes is known as **charging**.

Since the energy used in charging the cell can be looked upon as being stored until such time as it is taken from the cell by **discharging**, secondary cells are sometimes referred to as storage cells or accumulators. It is important to note, however, that they do not store electricity as such, and that there is no fundamental difference between the action of primary and secondary cells in supplying current.

Polarization

We have seen that when the simple cell supplies current, hydrogen collects at the copper electrode. In practice the electrode soon becomes covered with hydrogen bubbles, and is then kept more or less out of contact with the electrolyte. The result is that the current rapidly diminishes. A cell in this state is said to be **polarized**.

The defect is overcome in practical cells by the use in the cell of a substance which will react chemically with the hydrogen to form some harmless compound, thus avoiding the production of the gaseous layer. A substance able to perform this function is known as a **depolarizer**.

Local Action

It is difficult to obtain very pure zinc, and any foreign particles in the zinc electrode which come into contact with the electrolyte tend to form miniature cells, the particle of impurity forming one electrode and the surrounding zinc the other. The result is that a local current flows and the zinc is consumed even when the external circuit of the cell is not completed.

This unnecessary consumption of zinc can be largely avoided by amalgamating the zinc, *i.e.*, coating its surface with mercury. The mercury dissolves the outer layer of zinc and allows it to come into contact with the electrolyte, while acting as a screen to the impurities, which fall harmlessly to the bottom of the cell as the zinc is consumed.

Practical Primary Cells

We shall describe two practical forms of primary cell. The first is termed the **Daniell** cell, after John Frederic Daniell, an English chemist.

In this cell the zinc, copper and sulphuric acid are retained, but the acid is confined to a pot which surrounds the zinc electrode. The copper electrode is outside the pot and stands in a solution of copper sulphate, which acts as a depolarizer.

The pot is porous enough to allow some contact

between the acid inside and the copper-sulphate solution outside, so that the electrical action is not seriously impeded. When, however, the hydrogen appears at the copper electrode, it replaces some of the copper in the copper sulphate solution to form sulphuric acid, and instead of the hydrogen the displaced copper is deposited on the electrode. The e.m.f. of the Daniell cell is rather more than one volt.

The other primary cell which we will consider was invented by Georges Leclanché and named after him. In this the electrodes are a zinc rod and a carbon plate. The electrolyte is a solution of ammonium chloride (sal-ammoniac), a chemical compound of nitrogen, hydrogen and chlorine.

The depolarizer is powdered manganese dioxide, a chemical compound of manganese and oxygen. It is kept in position by a porous pot surrounding the carbon electrode. The pot does not prevent the electrolyte from soaking right through to the carbon.

When the cell supplies current, some of the nitrogen and hydrogen in the electrolyte is replaced by zinc from the zinc electrode. Part of this nitrogen and hydrogen forms ammonia, but there is left some hydrogen which would soon cause polarization were it not for the manganese dioxide, which parts with some of its oxygen to combine with the hydrogen in the formation of water.

The universally used " dry " cells are a special form of Leclanché cell. The zinc electrode forms the case of the cell, and next to it is a damp paste containing the ammonium chloride. The carbon electrode is in the centre and is surrounded by another damp paste containing the manganese dioxide.

Both in the " wet " and " dry " forms the Leclanché cell has an e.m.f. of about 1·5 volts.

Secondary Cells

Return again to Fig. 21, and this time suppose that both electrodes are of lead and that the electrolyte is again a solution of sulphuric acid. Since the electrodes are similar there is, of course, no e.m.f.

Now pass a current through the cell in the direction of the arrows. Oxygen from the electrolyte collects at electrode *a* and hydrogen at electrode *b*. The oxygen combines with some of the lead of electrode *a* to form a chemical compound known as lead dioxide, which in due course forms a layer all over the immersed electrode. The hydrogen is liberated.

We now have two dissimilar electrodes immersed in an electrolyte, and if we join them by an external conductor a current will flow in the direction opposite to that of the current previously passed through the cell.

Electrode *a* is termed the positive plate of the cell and electrode *b* the negative.

While the cell is being discharged, the lead dioxide at the positive plate and the metallic lead at the negative both become converted to lead sulphate, which is a chemical compound of lead, sulphur and oxygen. At the same time the strength of the electrolyte is gradually reduced by the formation of water.

To recharge the cell, current is again passed through it in the direction of the arrows. The result is to reconvert the lead sulphate to lead dioxide at the positive plate and to metallic lead at the negative. During the process the strength of the electrolyte is restored by the formation of acid.

A simple secondary cell of this kind would not supply very much current, and in practice it is necessary to treat the plates in order to give them a spongy structure offering a large surface to the electrolyte. It is possible to do this by passing current through the cell a number of times in alternate directions in order to " form " the plates, but it is now common practice to start, not with simple lead plates, but with lead (or lead alloy) grids filled with paste containing red lead for the positive plates and litharge for the negative. Both these materials are chemical compounds of lead and oxygen, and their use not only helps to shorten the manufacturing process but also enables the weight of the cell for a given output to be reduced.

Still further to increase the area of the electrodes, it is usual to form each one of several plates, the two sets being interleaved (without being allowed to touch) as indicated in Fig. 24. Separators of wood, glass or other material are used to keep the plates apart, while allowing the electrolyte to have free access to their surface.

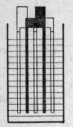

The e.m.f. of a fully charged cell is about 2·4 volts. During discharge this rapidly falls to about 2 volts, after which it remains fairly constant until the cell is nearly discharged, when it again falls rapidly. The cell should not be discharged after the e.m.f. has fallen to about 1·85 volts, as beyond this point the nature of the lead sulphate formed on the plates becomes such that it is not readily reconverted during the subsequent charging period. A cell which has been allowed to get into this condition is said to be sulphated.

FIG. 24.—Arrangement of Plates.

When, during charging, all the lead sulphate on the plates has been converted, hydrogen is given off freely from the negative plate and oxygen from the positive. The cell is then said to be gassing, and charging can be stopped.

The variation in the strength of the electrolyte caused by the formation of water during the discharging period and of acid during the charging period may be used as a guide to the state of a cell. Since the acid is heavier than water, the strength of the mixture can be measured by its specific gravity, *i.e.*, by the weight of a given volume as compared with the same volume of water.

Specific gravity is measured by an instrument known as a **hydrometer**. This is simply a weighted and graduated float (Fig. 25) which sinks farther into weak acid than into strong. The float is usually contained in a glass cylinder provided with a rubber suction bulb for drawing off a small quantity of the electrolyte for test. Do not confuse the words

FIG. 25.—Hydrometer Float.

" hydrometer " and " hygrometer "; the latter is the name of an instrument for measuring the degree of moisture present in the air.

The strength of acid commonly employed is such that the specific gravity when the cell is fully charged is about 1·2 and when fully discharged about 1·15. The specific gravity of pure sulphuric acid is 1·84 and that of water, of course, is 1. Hydrometers are often marked with four-figure numbers; 1200, for example, indicates a specific gravity of 1·2.

It is dangerous to pour water into strong sulphuric acid, and in mixing the two to make fresh electrolyte the acid must always be poured slowly into the water. It should not be necessary to add mixed electrolyte to a cell unless some has been spilt, since ordinary loss by evaporation is made up by adding water. The water should be distilled.

The capacity of a secondary cell for supplying current is measured in **ampere-hours**. A cell having a capacity of 10 ampere-hours will supply 1 ampere for 10 hours or, approximately, half an ampere for 20 hours, and so on. It is necessary to include the word " approximately " because the capacity in ampere-hours is greater for small rates of discharge than for large ones. For this reason it is usual to quote capacities " at the 10-hour rate," *i.e.*, on the assumption that the cell will be discharged in a period of 10 hours.

The lead-acid secondary cell is the type most commonly used, but it has a competitor in the alkaline cell, in which the active materials are based on nickel and either iron or cadmium. The electrolyte is a solution of caustic potash. The e.m.f. of this cell is only about 1·3 volts, and its efficiency (that is, the ratio of the amount of energy obtained during discharge to that put in during charge) is less than that of the lead-acid type. Its chief advantages are that it can be left in a discharged state without harm and that it can be charged and discharged at a high rate in proportion to its capacity in ampere-hours. These practices are likely to cause serious injury to a lead-acid cell.

Internal Resistance

We shall see in the next chapter that, in addition to passing from the positive pole to the negative pole over the external conductor, the current produced by a cell completes the circuit by passing from the negative pole to the positive pole through the cell itself. The **internal resistance** of the cell is therefore a matter of some importance.

Since the electrodes themselves are usually good conductors, the internal resistance is largely dependent upon the electrolyte. It will naturally be affected by the size of the electrodes and by their distance apart, since these are the factors which determine the effective cross-sectional area and length of the path through the electrolyte. In general, a large cell will have a smaller internal resistance than a small one. In primary cells the presence of a porous pot is likely to result in an increase in internal resistance.

The internal resistance of primary cells may be an ohm or more, according to their size and type. That of secondary cells varies greatly, according to their size, but it should not be more than a fraction of an ohm for a lead-acid cell. The internal resistance of alkaline cells is higher than that of equivalent lead-acid cells in good condition.

QUESTIONS

1. What is meant by electrolysis? Give an example.
2. How is electro-plating effected?
3. Name three practical applications of electro-chemical action.
4. Describe the action of a simple primary cell.
5. What is meant by (a) polarization and (b) local action in primary-cells? How can these faults be overcome?
6. What is the exact difference between a primary and a secondary cell?
7. Upon what does the voltage of a cell depend?
8. Describe briefly a practical type of primary cell.
9. What is a " dry " cell?
10. Explain the action of a secondary cell during charge and discharge.
11. In a lead-acid secondary cell, what variations occur in (a) the voltage and (b) the specific gravity of the electrolyte?
12. What is meant by the internal resistance of a cell and upon what does it depend?

CHAPTER V

THE ELECTRIC CIRCUIT

The Simple Circuit

WE are now in a position to examine the important idea conveyed by the term **electric circuit.**

Fig. 26 shows a resistance connected across the poles of a cell. The e.m.f. of the cell causes a current to flow through the resistance in the direction of the arrow.

The current does not suddenly appear at the positive pole and disappear at the negative, but passes round a continuous circuit, including the electrolyte between the two electrodes. The cell may, in fact, be looked

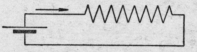

FIG. 26.—Simple Circuit.

upon as an electrical pump able to maintain the positive pole at a higher " level " (potential) than the negative, so that current flows from positive to negative in the external circuit.

Since, however, the cell has a definite internal resistance, some of the e.m.f. which it produces is required to drive the current through the cell itself. The potential difference between the positive and negative poles *when a current is flowing* is therefore less than the full e.m.f. of the cell.

As all parts of the circuit obey Ohm's law, it is easy to find what effect the internal resistance has upon the total current and the available voltage.

Example.—*A primary cell has an e.m.f. of 1·5 volts and an internal resistance of 2 ohms. What current will it send through a wire the resistance of which is 10 ohms?*

> *Internal resistance of cell* = 2 *ohms.*
> *Resistance of wire* = 10 *ohms.*
> *Total resistance* = 12 *ohms.*
> $Current = \dfrac{Voltage}{Resistance} = \dfrac{1·5}{12}$ *amp.* = 0·125 *amp.*

This is equivalent to a potential difference at the terminals of the cell of 1·25 volts, as compared with 1·5 volts when no current is flowing. The latter value is often spoken of as the voltage on open circuit and is, of course, equal to the e.m.f. of the cell.

Note that if the external circuit is broken at any point, a voltage equal to the full e.m.f. of the cell immediately appears across the gap. This will be clear if it is remembered that the flow of current ceases, so that there is no longer any voltage drop either in the internal or the external resistance, no matter how high these may be.

Since a large cell has a lower internal resistance than a small cell of the same type, it can send a larger current through any given external resistance, although the total e.m.f. is the same in both cases. The difference is most marked when the internal resistance forms an important part of the total resistance of the circuit— that is, when the external resistance is low.

Cells in Series

If two cells are connected in series, as shown on the left of Fig. 27, the positive pole of one being joined to the negative pole of the other, the total e.m.f. is the sum of those of the separate cells. This is obvious if we

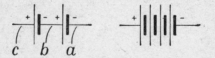

FIG. 27.—Cells in Series.

remember that the right-hand cell raises point *b* to a higher potential than point *a*, while the left-hand cell raises point *c* to a higher potential than point *b*.

Since the cells are in series, the internal resistances are also in series, and the total internal resistance is the sum of those of the separate cells.

Similar reasoning applies to any number of cells connected in series.

Example.—*Six cells each have an e.m.f. of 2 volts and an internal resistance of 0·5 ohm. Find the total e.m.f. and total internal resistance when they are connected in series.*

Total e.m.f. = (6 × 2) *volts* = 12 *volts.*
Total internal resistance = (6 × 0·5) *ohms* = 3 *ohms.*

In diagrams of cells connected in series, the connecting wires between adjacent cells are usually omitted, as shown for four cells on the right of Fig. 27.

Cells in Parallel

If two similar cells are connected in parallel, as in Fig. 28, the total e.m.f. is the same as that of one cell. Note that in this case the positive poles are connected together to form the joint positive pole, and the negative poles to form the joint negative pole. If the positive pole of each cell were connected to the negative pole of the other, the cells would quickly discharge through each other and no current would be sent to the external circuit.

Since the internal resistances are in parallel, the joint internal resistance is one half that of one cell. The

Fig. 28.—Cells in Parallel.

parallel connexion, in fact, is equivalent to the use of a single cell with larger electrodes.

Similar reasoning applies to any number of cells connected in parallel.

Example.—*Find the joint e.m.f. and internal resistance of the six cells in the last example when they are connected in parallel.*

E.M.F. = *e.m.f. of one cell* = 2 *volts.*
Internal resistance = (0·5 ÷ 6) *ohm* = 0·083 *ohm.*

Series and Parallel Groups of Cells

Cells can be arranged in series and parallel combinations. Thus, in Fig. 29 there are two parallel branches each of three cells in series. The total voltage is therefore three times that of a single cell. The total internal resistance *of each branch* is three times that of a single cell, and the joint internal resistance of the whole combination is half that of one branch, or $1\frac{1}{2}$ times that of a single cell.

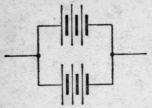

FIG. 29.—Series-Parallel Groups.

Note that it is not permissible to have more cells in one branch than another, since in this case the difference between the e.m.f. of one branch and the e.m.f. of the other would cause a current to circulate without any connexion to an external circuit. For the same reason, single cells of different voltages must not be connected in parallel.

If, as is usually the case, the external resistance through which it is required to pass current is high compared with the internal resistance of the cells, it will be found that the maximum current is obtained by connecting the cells in series.

Any group of two or more cells is called a **battery**. The same term is often loosely used for a single cell.

Example.—*In the battery shown in Fig. 29, each cell has an internal resistance of 1 ohm and an e.m.f. of 1·3 volts. If the battery supplies current to two 10-ohm resistances connected in parallel, what current will flow through each resistance ?*

Total e.m.f. = $(3 \times 1\cdot3)$ *volts* = $3\cdot9$ *volts.*

Internal resistance of 3 cells in series = (3×1) *ohms* = 3 *ohms.*

Internal resistance of complete battery = $\frac{3}{2}$ *ohms* = $1\cdot5$ *ohms.*

External resistance = $\frac{10}{2}$ *ohms* = 5 *ohms.*

Total resistance = $(1\cdot5 + 5)$ *ohms* = $6\cdot5$ *ohms.*

Total current = $\dfrac{e.m.f.}{resistance} = \dfrac{3\cdot9}{6\cdot5}$ *amp.* = $0\cdot6$ *amp.*

Current in each resistance = $\dfrac{0\cdot6}{2}$ *amp.* = $0\cdot3$ *amp.*

Circuit Diagrams

Where resistances occur in circuit diagrams, it is usual to represent them by the conventional sign and to ignore the resistance of the connecting wires. It should be remembered, however, that in practice the resistance of the connecting wires may have to be taken into account, particularly when the source of current is at some distance from the point at which the energy is used.

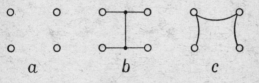

FIG. 30.—Methods of Connexion.

In some diagrams we have shown the junction of two conductors by a dot, but joints are avoided as far as possible in practical wiring. Suppose, for example, that it is required to join together the four points shown at *a* in Fig. 30. In a theoretical diagram we might represent this connexion in the manner shown at *b*, but in the actual wiring we should avoid the joints by adopting some such arrangement as that shown at *c*. Provided that the resistance of the connecting wires is negligible, the theoretical arrangement *b* and the practical arrangement *c* are identical from an electrical point of view.

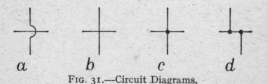

FIG. 31.—Circuit Diagrams.

When it is necessary to show conductors crossing each other without touching, the symbol shown at *a* in Fig. 31 is sometimes used, but the "bridge" is now often omitted, as shown at *b*. In the latter case it is desirable to avoid the connected crossing *c* to indicate wires in contact, and to represent such connexions in the manner shown at *d*.

Note that the use of the term "circuit" is not confined to a single closed loop. Almost any conductor or group of conductors is commonly called a circuit, from a single resistance to the complicated networks used in telephone and radio work.

Earth Returns

When it is required to complete a circuit from one place to another, it is sometimes possible to use the earth itself as a return conductor. The almost unlimited cross-sectional area of the path through which the current flows more than makes up for any deficiencies in con-

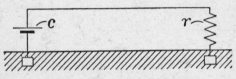

Fig. 32.—Earth-Return Circuit.

ductivity, and in practice the resistance of the earth, as distinct from that of the connexions to it, may be neglected.

Connexion is made by burying metal plates in damp positions, or by connexion to a water-pipe (not a gas-pipe) passing into the earth. Fig. 32 shows in diagrammatic form an earth-return circuit in which a cell *c* is supplying current to a distant device represented by the resistance *r*.

By analogy, the practice of using the metal body of a car or the metal hull of a ship as the return conductor in a lighting or other circuit is also spoken of as making use of an earth return.

Potentiometer

As an example of a circuit rather more complex than those so far considered, we will examine that of the **potentiometer**. This is an instrument the primary object of which is to compare or measure potential differences. It is based upon the same principle as the potential divider described on page 38.

In its simplest form the potentiometer consists of a uniform length of resistance wire (*i.e.*, wire made of one of the special alloys of high resistance mentioned in Chapter I), provided with a terminal at each end and a sliding contact able to make connexion at any point on its length. It is joined up as shown in Fig. 33, in which *b* is a battery able to send a steady current through the potentiometer wire *ae*, *s* the sliding contact, *g* a galvanometer, and *c* a cell under test.

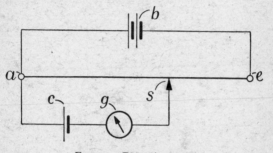

FIG. 33.—Potentiometer.

Consider the closed loop comprising cell *c*, galvanometer *g*, and the portion of the potentiometer wire between *a* and *s*. Since the wire is uniform, the voltage drop along any part of it is proportional to the length. The e.m.f. of the cell and the voltage between *a* and *s* are in opposition, and any current which flows through the galvanometer is dependent upon their *difference*.

Suppose that the voltage between *a* and *s* is less than that of cell *c*. Clearly, cell *c* is able to send a current round the circuit, and this will flow through the galvanometer from right to left. The galvanometer will be deflected accordingly.

Suppose now that the slider is moved to the right until the voltage between *a* and *s* is greater than that of the cell. The latter can no longer send a current, and instead, the voltage between *a* and *s* causes a current to flow through the galvanometer from left to right. The galvanometer is therefore deflected again, but this time in the opposite direction.

Between these two positions of the slider, a position can be found in which the e.m.f. of the cell is exactly balanced by the voltage drop between *a* and *s*. There is then no current through the galvanometer, and consequently no deflection.

In order to compare the e.m.f. of one cell with that of another, the first one is connected up as shown, and the slider is moved until the galvanometer shows no deflection. The position of the slider is then noted. Usually, there is a graduated scale for indicating its distance from the end of the wire.

The cell is next replaced by the second cell, and the slider is again moved until the galvanometer shows no deflection. The position of the slider is again noted. Then, since in each case the e.m.f. of the cell is proportional to the distance of the slider from the left-hand end of the potentiometer wire, the e.m.f. of either cell can easily be calculated if that of the other is known.

For laboratory use, a special type of cell which can be relied upon to maintain a constant e.m.f. is employed as a standard. The e.m.f. of any other cell (or any other potential difference) can then be readily determined.

Example.—*With a standard cell having an e.m.f. of 1·018 volts, a balance is obtained on the potentiometer in Fig. 33 when the slider is 0·51 metres from the left-hand end of the wire. When the standard cell is replaced by a Leclanché cell, a balance is obtained at 0·70 metres. What is the voltage of the Leclanché?*

The e.m.f. which balances the potential difference across 0·51 m of the potentiometer wire is 1·018 volts.

The e.m.f. which balances the potential difference across 0·70 m of the potentiometer wire must therefore be $(1·018 \times \frac{70}{51})$ volts = 1·4 volts.

Note that when a balance is obtained, no current is taken from the cell under test. The result is therefore the true e.m.f., and is not affected by the internal

resistance of the cell or the resistance of the galvanometer.

Note also that the method does not rely upon an
accurate galvanometer scale, since there is no deflection
when a balance is obtained. The galvanometer is not
being used for the *measurement* of current, but merely
to indicate whether or not a current is present.

The potentiometer can be used for many kinds of
electrical tests, but in all cases the action depends upon
the balancing of one potential difference against another.
In commercial forms, series resistances controlled by
switches are used in place of or in addition to the wire
and slider, but the principle is the same.

The Wheatstone Bridge

This is another testing circuit with which we should
be familiar. It is named after Sir Charles Wheatstone,
the pioneer of telegraphy, and is used for the comparison and measurement of resistances.

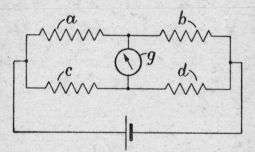

FIG. 34.—Wheatstone Bridge.

In Fig. 34, *a*, *b*, *c* and *d* are resistances and *g* is a
galvanometer. The cell shown in the lower part of
the figure sends a current through *a* and *b* and another
current through *c* and *d*. Since the two branches are in
parallel, the total potential difference is the same in
each case.

Suppose that *a* is equal to *b* and that *c* is equal to *d*.
Then the potential difference across *a* is equal to that
across *b*, and therefore equal to half the total.

Similarly, since the potential difference across c is equal to that across d, it too is equal to half the total. It follows that both sides of the galvanometer are at the same potential. There is therefore no current through it, and no deflection.

If we have a resistance (say d) the value of which we do not know, we connect it as shown, using a pair of equal resistances in the positions a and b. Then we try different resistances c until there is no deflection on the galvanometer. When we obtain a balance, we know that resistance d is equal in value to resistance c.

In practice, keys are inserted in series with the cell and galvanometer so that the circuit can be opened while the resistances are being changed.

If the resistance to be measured is fairly high or fairly low, it may be convenient to employ different values in positions a and b. Thus, if we know that d is fairly high, we may make the resistance of b ten times that of a. The potential difference across a will then be one eleventh of the total. A balance will therefore be obtained when the potential difference across c is also one eleventh of the total, *i.e.*, when it is one tenth of that across d. To find the value of d, we accordingly *multiply* the resistance of c by ten.

If, on the other hand, we know that d is fairly low, we may make the resistance of a ten times that of b. The potential difference across a will then be ten-elevenths of the total. A balance will therefore be obtained when the potential difference across c is also ten-elevenths of the total, *i.e.*, when it is ten times that across d. To find the value of d, we accordingly *divide* the resistance of c by ten.

Since resistances a and b serve to fix the ratio between the unknown resistance d and the resistance c with which it is compared, a and b are often called the ratio arms of the bridge. It is not necessary for their values to be accurately known, provided that their ratio is correct.

As a general statement of the relations of the four resistances, we may say that when a balance is obtained,

$$\frac{\text{Resistance of } a}{\text{Resistance of } b} = \frac{\text{Resistance of } c}{\text{Resistance of } d}$$

or Unknown resistance $d = c \times \dfrac{b}{a}$

Example.—*In the Wheatstone bridge circuit shown in Fig. 34, the value of resistance c which will produce a balance is 2·5 ohms and resistance a is one hundred times the value of resistance b. What is the value of the unknown resistance d ?*

Resistance $d = c \times \dfrac{b}{a} = (2 \cdot 5 \times 0 \cdot 01)$ *ohm* $= 0 \cdot 025$ *ohm.*

Wheatstone bridges are made in several forms. A very simple one consists of a wire and slider similar to that described for the potentiometer, the portion on one side of the slider serving for resistance a and the portion on the other for resistance b. In this case a fixed resistance is used for c, and the ratio of a to b (and therefore of c to d) is determined by the position of the slider when a balance is obtained.

More elaborate forms have groups of resistances controlled by switches or plugs so that the ratio of a to b and the value of c can be easily varied. One widely used pattern in which the resistances are controlled by plugs arranged to short-circuit those not in use is termed a **Post Office box.** (The term "to short circuit" means to bridge by a negligible resistance.)

Note that the Wheatstone bridge shares with the potentiometer the advantage of having no current passing through the galvanometer when a balance is obtained. The result is therefore independent of the resistance of the galvanometer or the accuracy of its scale.

QUESTIONS

1. Explain why the potential difference at the terminals of a cell supplying current is less than the e.m.f. of the cell.

2. A cell having an internal resistance of half an ohm sends a current of half an ampere through an external resistance of 1·5 ohms. What is the e.m.f. of the cell?

3. What current will two cells each having an e.m.f. of 1·5 volts and an internal resistance of 0·4 ohm send through an external resistance of 2·2 ohms if they are connected (a) in series, (b) in parallel?

4. Explain the principle of the potentiometer.

5. How can an unknown resistance be measured by means of a Wheatstone bridge circuit?

6. In using a potentiometer or a Wheatstone bridge, the aim is to obtain conditions which result in there being no deflection of the galvanometer. Why is this a good method of testing?

7. What is meant by the term *earth return circuit*?

8. The e.m.f. of a cell as measured on a potentiometer is 1·8 volts. When it is connected to an external resistance of 3 ohms, the current is found to be half an ampere. What is the internal resistance of the cell?

9. Through what external resistance will a cell having an e.m.f. of 2 volts and an internal resistance of one tenth of an ohm send a current of 5 amperes?

10. A battery consists of four cells connected in series. Each cell has an e.m.f. of 1 volt and an internal resistance of 1 ohm. What is the e.m.f. and internal resistance of the complete battery, and what current will it send through an external resistance of 16 ohms?

CHAPTER VI

HEAT AND ELECTRICAL ENERGY

Heating Effect of Current

Reference has already been made to the fact that the passage of current through a resistance results in the generation of heat. Heat is a form of energy, and the simplest approach to its understanding is to consider the amount of energy required to raise a given quantity of water from one temperature to another.

As an experiment, we might immerse a length of insulated wire of known resistance in a vessel containing a measured quantity of water, and find the rise in temperature of the water caused by passing a known current through the wire for a known time. Provided that we took proper precautions to avoid loss of heat, and allowed for the heat required to raise the temperature of the vessel itself, we should find that the electrical energy represented by 1 ampere flowing through a resistance of 1 ohm for 1 second was equivalent to the heat energy required to raise the temperature of 0·24 gramme of water by 1° C.

If we passed 1 ampere through two resistances, each of 1 ohm, connected in series, twice the heat would be produced. In general, the heat produced by a given current in a given time is proportional to the resistance through which it flows.

Suppose that the conductor we are considering is that shown on the left of Fig. 35, and that we split it lengthwise, as shown in the centre. Each half will carry half the current and produce half the heat of the original.

Each half, however, will be of twice the resistance of the whole, so let us imagine that each half is composed

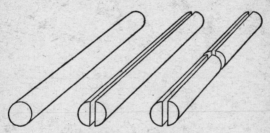

FIG. 35.—Division of Conductor.

of two quarters, as suggested on the right. Each quarter will then have the same resistance as the original, will carry half the current and will produce **one quarter** of the heat.

We conclude, therefore, that for a given resistance the heat produced is proportional, not to the current, but to the square of the current. In other words, twice the current will produce four times the heat, and so on. This is a most important relation.

To sum up, we may say that the heat due to a given current flowing through a given resistance for a given time is proportional to I^2Rt,

where I is the current in amperes,
R ,, ,, resistance in ohms,
t ,, ,, time in seconds.

Example.—*The heat produced by a current of 2 amperes flowing through a resistance of 125 ohms for 5 minutes is* *used to heat 1000 grammes of water. If no heat is lost, what will be the rise in temperature of the water?*

1 *ampere flowing through* 1 *ohm for* 1 *second raises* 0·24 *gramme of water by* 1° *C.*

Heat is proportional to I^2Rt.

Therefore rise in temperature is:

$$\frac{2^2 \times 125 \times 5 \times 60 \times 0·24}{1000} = 36° \ C.$$

Note that although the heat produced by a given current is proportional to the resistance through which the current flows, the current taken from a supply of constant voltage goes down if the resistance is increased. As the heat is proportional to the square of the current, the increase in resistance is more than counterbalanced by the fall in current, and the net result is a decrease in heat.

Thus, doubling the resistance would double the heat if the current remained the same. But if the voltage is constant, doubling the resistance will halve the current. Halving the current would normally divide the heat by four, and even with the doubled resistance the heat will be halved.

Applications of the Heating Effect

The heating effect of a current can be applied to almost any situation in which heat may be required. Special alloys which combine high resistivity with the ability to withstand high temperatures are used for the resistances in which the heat is produced. They may be designed to operate at any temperature up to red heat.

In order to obtain a high enough resistance, it is usually necessary to employ a length of the resistance wire in the form of a spiral; this is mounted on a heat-resisting support, and called a **heating element.** Electric heaters, ovens, kettles, irons, and similar devices make use of the heating effect of the current in this manner.

In some electric furnaces, metal is maintained in a molten state by passing a very heavy current through the metal itself.

Electric **resistance welding** is also dependent upon the direct heating effect of the current. In **butt welding**, the edges of the metal plates to be joined are forced together and a heavy current is passed from one to the other. As most of the resistance in the path is at the joint (where the contact is never perfect), the metal here becomes hot enough for the two parts to weld together, while the remainder remains comparatively cool.

In **spot welding**, the faces of two sheets of metal are placed together and gripped between two metal electrodes, as shown in Fig. 36. Current is passed from one electrode to the other through the two thicknesses

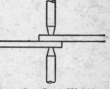

FIG. 36.—Spot Welding.

of metal. Considerable local heat is thus produced in the part of the metal lying between the electrodes, and the two sheets are welded together at this point.

Electric Filament Lamps

The elements used in ordinary electric heaters cannot be raised to more than red heat without being rapidly destroyed by oxidation (combination with the oxygen in the air). A conductor operated in a vacuum, however, can be safely raised to a white heat, and in this state will emit a considerable amount of light. This is the basis of the ordinary filament lamp.

Early lamps had filaments of carbon, and this type is still occasionally employed for situations in which vibration is present. For ordinary use, however, they have long been superseded by lamps with metal filaments, which will produce several times the amount of light for the same consumption of energy. The metal now used for the filaments is tungsten.

Even in the absence of oxygen, the life of a lamp filament is limited, particularly if it is run at the high temperatures which are desirable from the point of view of efficiency. The allowable temperature can be increased by filling the glass envelope with an inert gas such as argon. Lamps in which this is done are said to be **gas-filled**. They can be made to give up to nearly twice the light of a vacuum lamp for the same consumption.

It is easy to increase the temperature at which a lamp filament is run by increasing the voltage beyond that for which it is designed. The result is to improve the light-giving efficiency, but to decrease the life of the lamp to an uneconomical extent.

Note that as the resistance of a lamp filament varies, like that of most other conductors, according to the temperature, its " cold " resistance cannot be used to calculate the current it will take under working conditions.

Current-Carrying Capacity

Since all conductors offer resistance, they become heated to a greater or smaller extent when they carry current. The resistance of conductors used to convey current from one point to another must therefore be small enough to ensure that they do not become so hot as to injure their insulation or to cause danger of fire.

It must also, of course, be small enough to ensure that the voltage drop in the conductors is not unduly large.

The voltage drop can be calculated from the current and resistance. The safe current from a heating point of view has been laid down for various types and sizes of cables in tables issued by the Institution of Electrical Engineers.

Fuses

There is always a possibility that too low a resistance will be connected across an electricity supply, thus drawing a current which would cause overheating of the connecting wires. An extreme but very common case is an accidental short circuit. As a safeguard, fuses are inserted.

A fuse is simply a " weak link " in a conductor, designed to break the circuit by melting before the current becomes large enough to cause any danger. Fuse-wire is usually made from tin–lead alloy for small currents and from tinned copper for larger ones.

The current-carrying capacity of fuses is dependent upon their size, the metal of which they are made, and the manner in which they are mounted. In **cartridge fuses** the fuse-wire is enclosed in a tube filled with incombustible material.

The conventional way of indicating a fuse in a circuit diagram is shown in Fig. 37.

It is not practicable to design a fuse which will break a circuit on a very small overload, since there is necessarily a fairly wide margin between the current that it will carry safely and the current that is certain to melt it.

FIG. 37.—Fuse.

In any case, it would seldom be desirable to open a circuit on account of a small momentary overload. In telephone line circuits and some others, a device called a **heat coil** is employed to deal with slight but persistent overloads or stray currents. This consists of a small coil of resistance wire which, if an excess current persists for more than a short time, melts a soldered joint and allows a spring to open the circuit.

Direct Production of Current by Heat

Fig. 38 (left) shows strips of two dissimilar metals, say iron and copper, joined together at each end. We

FIG. 38.—Thermo-Couple.

will suppose that the upper strip is iron and the lower one copper.

If the junction on the left is maintained at a moderately higher temperature than the one on the right, a current is found to flow in the direction of the arrow. The effect is still evident if the other joint is opened to admit the connexion of a galvanometer in the circuit, as indicated at *g* on the right of the figure.

An arrangement of this kind is termed a **thermo-couple**. Although the e.m.f. from a single junction is only a fraction of a volt, and the current correspondingly low, the device has several commercial uses. Thus, it is employed in the **thermo-electric pyrometer,** which, like the resistance pyrometer mentioned on page 26, is used for the measurement of temperatures too high for an ordinary thermometer. Thermo-couples are used also in some kinds of electrical measuring instruments (Chapter XVI).

Energy and Power

We have dealt, so far, with the following units:

Coulomb (unit of quantity).
Ampere   (,, ,, rate of flow, or current).
Ohm   (,, ,, resistance).
Volt   (,, ,, potential difference and e.m.f.)

In addition, we have mentioned electrical **energy.**

Energy is a measure of the capacity to do work, and may take many forms. In mechanics, for example, a definite amount of energy is required to lift a given weight to a given height.

The modern unit of energy is the **joule,** named after the English physicist James Prescott Joule. In electrical work it represents the energy expended when one coulomb passes through a conductor across which the potential difference is one volt. We may therefore write:

$$\text{Joules} = \text{Volts} \times \text{Coulombs.}$$

Power is the rate of doing work. The old term horse-power, for instance, was defined by the weight that a standard " horse " was supposed to be capable of lifting to a given height in a given time.

The modern unit of power is the **watt,** which takes its name from James Watt, the engineer. Although it began as a purely electrical unit, it has now come into more general use.

One watt represents a rate of one joule per second, and we may therefore write:

$$\text{Watts} = \frac{\text{Joules}}{\text{Seconds}}.$$

But since

$$\text{Joules} = \text{Volts} \times \text{Coulombs},$$

it follows that

$$\text{Watts} = \frac{\text{Volts} \times \text{Coulombs}}{\text{Seconds}}.$$

We have seen, however, that

$$\frac{\text{Coulombs}}{\text{Seconds}} = \text{Amperes},$$

so that we have finally,

$$\text{Watts} = \text{Volts} \times \text{Amperes}.$$

This is an important practical relation. It shows us that the power in a circuit is dependent simply upon the voltage and the current.

Example.—*What current is taken by a 100-watt lamp connected to 250-volt mains ?*

100 (*watts*) = *current* (*amps*) × 250 (*volts*).
Current = $\frac{100}{250}$ *amp* = 0·4 *amp*.

The watt is inconveniently small for many purposes, so a multiple unit, the **kilowatt,** is also used. One kilowatt (abbreviated kW) is equal to 1000 watts.

In symbolic form, the relation between power, current and voltage is given by:

$$P \text{ (power)} = V \text{ (voltage)} \times I \text{ (current)},$$

and since from Ohm's law (page 33),

$$V = IR,$$

it follows that

$$P = IR \times I = I^2R.$$

Electrical power (and therefore electrical energy, which is power × time) is thus proportional to the square of the current and to the resistance. This confirms the result we obtained at the beginning of this chapter for the amount of heat energy into which electrical energy can be converted.

We can now add to our list of units:

Joule (unit of energy).
Watt (,,   ,, power).

We have seen that heat is proportional to I^2Rt. This may be written $I(IR)t$, and is therefore equivalent to:

$$\text{Amps} \times \text{Volts} \times \text{Seconds},$$

that is, to watt-seconds or joules.

The joule is therefore the natural unit in which to measure heat energy. It has replaced the calorie, which is the heat required to raise the temperature of 1 gramme of water by 1° C. From what was said on page 62, it follows that one joule is equal to 0·24 calorie.

Note that, although the watt is the unit of power, the term horse-power has not gone out of use. One horse-power is equal to 746 watts.

Payment for Electrical Energy

When electricity is supplied, say, to a private house, it is energy, rather than current, which is used, and for which payment is required. Electrical energy can be measured, as we have seen, in joules (or watt-seconds), but the joule is an inconveniently small unit for the present purpose and so a larger one, the **kilowatt-hour,** is actually employed. The kilowatt-hour is the energy represented by 1 kilowatt taken for 1 hour; it is therefore equal to 3,600,000 watt-seconds or joules.

Sometimes the charge for energy is based solely upon the number of units (kilowatt-hours) taken, but there may also be a standing charge determined by the size of the consumer's premises or by the maximum power demanded at any one time.

QUESTIONS

1. In what units is heat energy measured?
2. How does the heating effect of a current vary with (a) the strength of the current and (b) the resistance through which it flows?
3. How many times the heat produced in a 150-ohm resistance will be produced in a 50-ohm resistance connected to the same voltage supply?

4. Give three examples of the application of the heating effect of a current.

5. What is meant by (a) a vacuum lamp and (b) a gas-filled lamp ?   What is the advantage of the latter ?

6. What considerations limit the amount of current which can be passed through a conductor ?

7. What is the object of a fuse ?

8. What is a thermo-couple ?

9. What units are used to measure (a) electric energy and (b) electric power ?

10. State the relation between the units in each of the following groups :—

(a) Volts, amperes, watts.
(b) Volts, coulombs, joules.

11. An electric kettle is marked " 720 watts, 240 volts ". Find (i) the current, (ii) the number of units per hour which it is intended to take.

12. How many hours' light will a 25-watt lamp give on one unit of electricity?

13. What is the value in ohms of a resistance which takes 60 watts when connected to 240-volt supply mains?

14. If electrical energy costs two pence per unit, how much per hour does it cost to run a 100-watt lamp?

15. What is the cost of raising the temperature of a litre of water by 60° C if electrical energy costs one penny per unit and 20% of it is wasted?   (One litre of water weighs 1000 grammes.)

CHAPTER VII

MAGNETS AND MAGNETIC MATERIALS

SOME simple facts about magnetism were mentioned in Chapter I.   It is now time to examine this subject more closely.

Magnetic Substances

The ordinary permanent magnet (that is, one that will retain its magnetism without electrical aid) is made of steel, which is an alloy of iron and carbon.   Iron is the most important magnetic material, but appreciable effects can also be obtained from nickel and cobalt. All these substances (iron, steel, nickel and cobalt) are termed **ferromagnetic,** and are attracted by a magnet.

Most other substances are practically non-magnetic. They may, however, be divided into two classes: the **paramagnetic** materials, which have a faint tendency to be attracted by a magnet, and the **diamagnetic** materials, which have a faint tendency to be repelled. These effects are so slight that they may be ignored for practical purposes, so that when we speak of a magnetic material we mean one that is ferromagnetic, usually iron or one of its alloys.

Magnetic Field

Reference has already been made to the magnetic field surrounding a magnet, and to the representation of this field by lines (often called lines of force) such as those in Fig. 1 (page 13). There are, of course, no gaps between the lines of force, which are nothing more than indications of the direction of the field at particular places. As the field cannot have two directions at the same place, it is impossible for two of the lines of force to cross each other.

The attractive force of a magnet is most marked at two points near its ends, termed its north and south poles. The tendency of a compass needle placed near a magnet to set itself in the direction of the field may be interpreted as being due to the combined effect of the poles of the magnet upon the poles of the compass needle, each pole of the needle being attracted by the " unlike " pole of the magnet and repelled by the " like " pole. We may sum this up by saying that unlike poles (a north and a south) attract each other, while like poles (two norths or two souths) repel each other.

Since unlike poles are attracted, we ought, strictly speaking, to use the name " north pole " for the one which, in the case of a freely suspended magnet, points to the *south* magnetic pole of the earth. This is not done, and by the north pole of a magnet we mean the one that tends to point towards the north. As a concession to accuracy, the term " north-seeking " is sometimes used, but it has never achieved much popularity.

When a magnet (Fig. 39, left) acts on an unmagnetized piece of iron (Fig. 39, right), magnetism is induced in the iron, the end nearer the magnet acquiring a polarity such that it is attracted by the neighbouring pole, and the other end, of course, the opposite polarity. This explains not only why a previously unmagnetized piece of iron is attracted, but also why the pole of a strong magnet will attract even the like pole of a weak

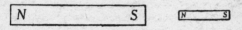

FIG. 39.—Induced Magnetic Poles.

one. In the latter case the magnetism of the weak magnet is reversed by the magnetism induced by the strong magnet, and attraction takes place.

A magnetic body arranged to be attracted by a magnet is termed an **armature**.

Electromagnets

We have seen that a coil of wire in the form of a solenoid has, while it is carrying current, a magnetic field similar to that of a bar magnet, and that it will

FIG. 40.—Electromagnet.

magnetize a bar of iron placed within it. A bar of iron surrounded by a magnetizing coil in this way forms an **electromagnet** and is represented in its simplest form in Fig. 40.

The question arises as to which end of the iron becomes a north pole and which a south. This can be worked out by considering the direction of the field produced by the turns of the coil, but there is a simpler way.

Look at one end of the magnet. If the direction of the current in the coil is then the Same as that of the hands of a clock (Fig. 41, right), the end being examined is the South pole. If it is Not the same (Fig. 41, left), the end is the North pole.

FIG. 41.—Poles of Electromagnet.

In Fig. 40, for example, the left-hand end is the north pole and the right-hand end the south.

Note that the direction in which the current progresses along the length of the magnet in passing round successive turns is immaterial. All that matters is the direction in which the current moves *round* the magnet.

Electromagnets are often made of "horseshoe" shape, as shown in Fig. 42. The two poles are thereby

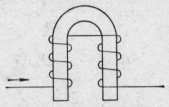

Fig. 42.—Horseshoe Magnet.

brought fairly close together, and instead of being spread over long paths in the air, most of the lines of force are concentrated between the poles. An armature arranged to bridge the poles is very strongly attracted, and magnets of this type can be made to support heavy weights.

Note that the two limbs of the horseshoe magnet appear at first sight to be wound in opposite directions. If, however, the magnet is imagined to be straightened out into the bar form, it will be seen that this is not really the case.

Uses of Electromagnets

Since electromagnets are so familiar, we shall mention only a few examples of their uses. In the electric trembler bell the attraction of an armature by the magnet is arranged to open a contact in the magnet circuit, so that the armature falls back to its original position. The contact is thus reclosed and the armature again attracted, the operation continuing in this manner so long as the external circuit remains closed.

Electromagnets are used in many kinds of signalling apparatus, and in relays for opening or closing one circuit when current is received over another. In the telephone receiver, variations of the current in a magnet winding are made to vibrate a thin iron diaphragm, and thus to reproduce the vibrations at a distant transmitter which caused the variations in current.

Larger electromagnets are employed in magnetic clutches for coupling one rotating shaft to another at will, while both electromagnets and permanent magnets are used in magnetic chucks for holding magnetic material while it is being operated upon by machine tools.

The Magnetic Circuit

Consider the case of an iron ring wound as shown in Fig. 43. A winding of this kind is known as a **toroidal**

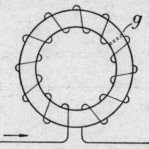

FIG. 43.—Toroidal Coil.

coil. When current flows in the winding, a **magnetic flux** is set up in the iron core, and the closed path taken by the flux is termed the **magnetic circuit.**

Since lines of force assume the form of closed loops, there is a similar magnetic circuit in other cases. In a straight bar magnet, for example, part of the circuit consists of the magnet itself, and part of the surrounding space.

The properties of the magnetic circuit can be described by comparing it to an electric circuit, but it is important to remember that in the magnetic circuit there is nothing in the nature of a flow of current. In an electric circuit we have, according to Ohm's law,

$$\text{Current} = \frac{\text{Electromotive force}}{\text{Resistance}}.$$

In the magnetic circuit we have flux instead of current and a property called **reluctance** instead of resistance. The magnetic quantity corresponding to electromotive force is called **magnetomotive force** or **m.m.f.**, and we may write:

$$\text{Flux} = \frac{\text{Magnetomotive force}}{\text{Reluctance}}.$$

Earlier units of magnetic flux have been replaced by the **weber***, named after the German physicist Wilhelm Eduard Weber. An equally important quantity for many purposes is the **flux density**; thus in the toroidal coil in Fig. 43 we should be interested not only in the total flux, but also in its density over the cross section of the core. The unit of flux density is the **tesla**, which is 1 weber per square metre.

The magnetomotive force is produced by the magnetizing winding. It is proportional to the number of turns and to the current flowing round them, *i.e.*, to the number of **ampere-turns**. It does not matter, for example, whether we have 1 ampere flowing round 1000 turns or half an ampere flowing round 2000 turns; in both cases there are 1000 ampere-turns and the magnetic effect is the same.

The reluctance is proportional to the length and inversely proportional to the cross-sectional area of the magnetic path, just as electrical resistance is proportional to the length and inversely proportional to the cross-sectional area of a conductor. Like electrical resistance, also, reluctance depends upon the material of which the path is made.

When we say that one material is a better electrical conductor than another, we mean that it has a lower resistivity. It would, however, be equally correct to say that it had a higher conductivity. In the magnetic case this course is taken, and the term **permeability** is used for the magnetic equivalent of conductivity.

If the permeability of a perfectly non-magnetic substance is taken as unity, that of a paramagnetic substance is very slightly greater than unity and that

* Pronounced *vayber*.

of a diamagnetic substance very slightly less. We may, however, ignore these small differences.

Ferromagnetic substances, on the other hand, may have a permeability several thousand times as great as that of a non-magnetic substance. If we say, for example, that the sample of iron forming the core of the toroidal coil we have been considering has a relative permeability of 1000, we mean that under the prevailing conditions the flux produced is 1000 times as great as it would be with an air core.

There is this important difference between magnetic permeability and electrical conductivity: whereas, apart from temperature variations, electrical conductivity is constant for a given material, the permeability of

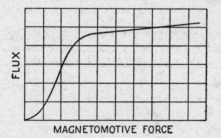

FLUX

MAGNETOMOTIVE FORCE

Fig. 44.—Magnetization of Iron.

magnetic materials varies according to the strength of the magnetic field.

In Fig. 17 (page 35) we drew a graph showing the relation between current and voltage, and the result was a straight line. A similar graph drawn to show the relation between flux and magnetomotive force in a magnetic circuit containing iron takes the general form shown in Fig. 44, in which vertical heights represent flux and horizontal distances magnetomotive force.

If no iron had been present, the flux, although much smaller, would have been proportional to the magnetomotive force, and the result would have been a straight line. The effect of the iron is slight in the early part of the curve, then it causes a very rapid increase in flux for small increases in magnetomotive force, and finally

it is again slight. When the last stage is reached, the
iron is said to be saturated.

It follows, from this variation, that we cannot state
a definite permeability for a given magnetic material.
Fig. 45, in which vertical heights represent permeability
and horizontal distances magnetomotive force, is a
permeability curve corresponding to Fig. 44.

In practice, curves relating flux density (teslas, or
webers per square metre) to m.m.f. per unit length
(ampere-turns per metre) are plotted experimentally for

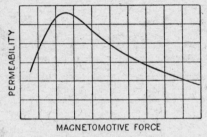

FIG. 45.—Permeability Curve.

each material. From them and the dimensions of the
core the necessary ampere-turns can be determined.

Magnetic circuits are usually more complex than the
simple ring we have been considering, but in most cases
it is possible to split them into separate portions, the
reluctance of each of which can be estimated with a fair
degree of accuracy. Thus, if the iron ring in Fig. 43
had a gap at *g*, we should find the reluctance of the iron
path and that of the air-gap and add the two together
to obtain the total reluctance. As the permeability
of iron is much greater than that of air, we should
probably find that the gap was responsible for the greater
part of the total.

Magnetic Leakage

We have seen that an appreciable magnetic field can
be set up in a non-magnetic substance, and the same is
true even of a vacuum. There is no magnetic " insulator " by which we can confine the flux to well-

defined paths. Magnetic calculations are therefore complicated by uncertainty as to how much flux will take the path provided and how much will spread where it is not wanted. Owing to this leakage factor, calculations are usually less exact for magnetic than for electric circuits.

The lack of a magnetic insulator is felt also when it is desired to keep magnetic flux away from a particular area. It is often possible, however, to provide a path of low reluctance which will take most of the flux, and this principle is used in magnetic screens for protecting delicate apparatus from unwanted fields.

Hysteresis

The dotted line in Fig. 46 corresponds to the first part of the curve in Fig. 44 and represents the increase

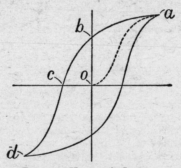

FIG. 46.—Hysteresis Curve.

in flux in an iron core as the current in a magnetizing winding is gradually increased. Vertical heights represent flux, and horizontal distances represent ampere-turns.

If, when the point a is reached, the current is gradually reduced again to zero, the flux, instead of also returning to zero, follows the full-line curve to b, thus showing that after the magnetizing force has been removed there is still some residual magnetism left in the iron. The power of retaining magnetism in this way is known as retentivity or remanence and is related to the height ob in the diagram.

If the connexions of the magnetizing coil are now reversed and the current is again gradually increased, it tends to magnetize the core in the opposite direction. At length the residual magnetism is overcome, and at the point *c* the iron is no longer magnetized. The distance *oc* is a measure of the force necessary to destroy the residual magnetism. The power of resisting demagnetization in this manner varies greatly with the material and is termed **coercivity**.

If the reversed current is further increased until a point *d* is reached, and then gradually reduced to zero and again increased in the original direction, the complete full-line curve will be obtained. It is evident that the changes in magnetization always lag behind the changes in magnetizing force which produce them, and this tendency is termed **hysteresis**.

The complete full-line figure is known as the hysteresis loop, and its area is a measure of the energy lost in changing the magnetic state of the material employed. This energy appears as heat in the core. It is greatest when the hysteresis loop is wide, as in the case of hard steel, and least when the loop is narrow, as in the case of soft iron.

High retentivity and high coercivity are clearly desirable in material for a permanent magnet. On the other hand, these properties are not wanted in an electromagnet which is required to lose its magnetism as soon as the current is removed.

A permanent magnet is demagnetized if it is raised to a red heat, while mechanical vibration is also detrimental. Even the passage of time results in appreciable weakening, and magnets the strength of which is required to remain constant are artificially " aged ".

Magnetic Materials

Soft iron has good permeability, low retentivity and low coercivity, and is therefore suitable for electromagnets. The addition of a small proportion of silicon produces an alloy with a lower hysteresis loss and a higher ohmic resistance. By the last quality we mean a higher resistance when considered as an electrical conductor;

we shall see in a later chapter that this is a desirable feature in many cases.

Alloys of iron and nickel have some interesting properties. Although both of these are magnetic materials, it is possible to make from them an alloy which is practically non-magnetic. Combined in other proportions they form materials having low hysteresis losses and high permeability, especially at low magnetizing forces. In most cases special heat treatment is necessary to obtain the desired characteristics.

Hard steels, particularly those containing tungsten, chromium or cobalt, have high coercivity and are suitable for permanent magnets. Certain alloys of iron, nickel, aluminium and cobalt are still better.

As a final example of the curious variations in magnetic properties produced by alloying different materials, we may mention that magnetic alloys of copper, aluminium and manganese are known, although all these substances are practically non-magnetic.

QUESTIONS

1. What is meant by the statement that a substance is (a) ferromagnetic, (b) paramagnetic, (c) diamagnetic?

2. If "like poles repel", why will the north pole of a strong magnet attract the north pole of a weak one?

3. How can the polarity of an electromagnet be determined from the direction of current in the winding?

4. What is an ampere-turn?

5. What quantities in the magnetic circuit are comparable with (a) current, (b) electromotive force, and (c) resistance in the electric circuit? How are the magnetic quantities related?

6. What is meant by reluctance, and upon what does its value depend?

7. What do you understand by the permeability of a magnetic substance?

8. Has the permeability of a given substance always the same value? If not, how does it vary?

9. What is meant by hysteresis?

10. State the magnetic qualities desirable in (a) a permanent magnet and (b) an electromagnet, giving in each case examples of materials in which they are to be found.

11. The winding of an electromagnet carries a current of half an ampere. Find the current necessary to give the

same magnetic effect if the number of turns is increased by 25%.

12. The resistance of the winding of an electromagnet is 10 ohms, and the p.d. across the winding is 5 volts. If half the turns are removed, and the resistance is thus reduced to 5 ohms, what voltage is necessary to produce the same magnetic effect as before?

13. A certain winding is to have 5000 ampere-turns, its resistance is to be 100 ohms, and it is to be operated on 250-volt supply mains. How many turns are necessary?

CHAPTER VIII

ELECTRICAL AND MECHANICAL ENERGY

Electromagnetic Induction

FIG. 47 shows a solenoid wound on an insulating tube so that a bar magnet can be inserted from one end. The winding is connected to a galvanometer *g*.

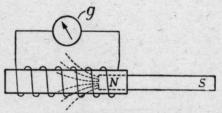

FIG. 47.—Electromagnetic Induction.

With this apparatus we can demonstrate some of the relations between electrical and mechanical energy. If the magnet is thrust quickly into the coil, north pole first, the galvanometer is deflected, showing that a current has been produced. As soon as the movement of the magnet stops, the flow of current stops too.

If the magnet is now quickly returned to its original position, a current again flows, but this time in the opposite direction. If the south pole of the magnet is inserted, the direction of the current is the same as when the north pole is withdrawn, and vice versa.

These effects are said to be due to **electromagnetic**

Induction, and their cause is to be found in the field by which the magnet is surrounded. Part of this field is shown in the diagram by dotted lines. As the magnet is moved, the magnetic flux passing through each turn of the coil is altered. This change in **flux linkage** produces an e.m.f. across each turn, and as the turns are in series, their effects are added together. The result is that a current flows in the galvanometer circuit while the magnet is moving.

If the north pole of the magnet is inserted, the current flows in the direction that will produce a north pole at the mouth of the coil, thus opposing the insertion. If the north pole is withdrawn, the current flows in the direction that will produce a south pole at the mouth of the coil, thus opposing the withdrawal. In all cases the effect of the current is to oppose the change which produces it, or to try to maintain the magnetic field in its original state. This is known as Lenz's law.

Generator Effect

Suppose that the rectangle N in Fig. 48 is an end view of the north pole of a bar magnet, the dots representing

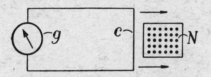

FIG. 48.—Change in Flux Linkage.

the flux emerging from it. If a conductor c, connected to a galvanometer g, is moved from left to right close to the pole, as indicated by the arrows, the amount of flux threading the circuit will change, and again a current will flow while the change continues. A similar effect would be produced if the magnet moved from right to left while the conductor remained stationary.

It is usual to think of the conductor as *cutting* the flux, and provided that we do not forget that the essential thing is the change in flux linkage, this is quite permissible. The e.m.f. produced is then proportional to the rate of cutting flux.

The relation between the units is such that an e.m.f. of 1 volt is produced for every weber cut per second. The weber, as we saw in the last chapter, is the unit of flux. To find the number of webers cut per second, we first take the useful portion of the conductor (*i.e.*, the portion which is actually moving across the field), and multiply its length in metres by its speed in metres per second. This gives the area swept out per second in square metres, and this area, multiplied by the flux density, gives the number of webers.

Example.—What e.m.f. is produced in a conductor 20 centimetres long moving at the rate of 1000 centimetres per second across a field of 0·8 tesla?

Area swept out in one second = (20 × 1000) *square centimetres*
= 2 *square metres.*
Flux cut per second = 2 × 0·8 = 1·6 *webers*
∴ *E.M.F.* = 1·6 *volts.*

We must now consider the direction of the induced current. Fig. 49 (left) shows another view of a magnetic

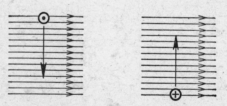

FIG. 49.—Conductor cutting Field.

field and moving conductor. The conductor is represented in cross section by the small circle, and the direction of the field by arrow-heads. The latter point, as usual, to the south pole of the magnet that is producing the field.

If the conductor is moved in the direction of the arrow, the induced e.m.f. will cause a current to flow (provided, of course, that the circuit is closed) *up* out of the paper, as indicated by the dot on the conductor. If the conductor is moved in the opposite direction, as shown on the right of the same figure, the current will flow *down* into the paper, as indicated by the cross.

Note that if either the direction of the motion or the direction of the field is reversed, the direction of the current is reversed also, but that if the directions of both the motion and the field are reversed, the direction of the current remains the same.

Note also that if the movement is not at right angles to the field, the amount of flux cut for a given movement is reduced. The e.m.f. is therefore smaller if the speed remains the same.

The direction of the current can be remembered by what is known as the "right-hand rule". It will be found that the thumb, forefinger and middle finger can easily be extended so that each is at right angles to the other two. If the thuMb (of the right hand) then points in the direction of the Motion, and the First finger in the direction of the Flux, the seCond finger will point in the direction in which the Current flows. The last two fingers of the hand should be bent down in order to avoid confusion.

The conductor moving in the magnetic field is a simple example of the conversion of mechanical to electrical energy. Machines for doing this are termed **dynamos** or **generators**.

Elementary Generator

The conductor in Fig. 49 will soon pass right through the magnetic field, when the current will cease. As a further step, therefore, let us take a complete loop of wire and mount it so that it can be rotated in the field, as shown in Fig. 50, in the upper part of which the wire loop is shown at *w* and the axis about which it is rotated at *x y*. It is assumed that the field is produced by a horseshoe magnet, shaped so that its poles *N* and *S* face one another.

The lower part of the figure is a cross section showing the two sides of the wire loop in the magnetic field. The arrow indicates the direction of rotation.

When the loop is in the position shown, both sides of it are cutting lines of force at the maximum rate, and an application of the right-hand rule (or a comparison with Fig. 49) will show that the e.m.f. induced in one side assists that induced in the other in sending current in the direction indicated.

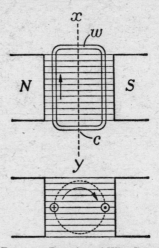

FIG. 50.—Rotation of Wire Loop.

When the conductors forming the sides of the loop have moved forward 90° (one quarter of the dotted circle) from the position shown, they are no longer cutting lines of force, but are " sliding between them ", and for a moment there is no e.m.f. As the rotation continues, they start to cut the field again, but each in the opposite direction. During each complete revolution, therefore, current flows first in one direction round the loop, then dies away, then flows in the other direction, and then dies away again.

We are now faced by two problems: the first is to correct this continuous reversal of current, and the second is to get the current out of the loop and into stationary conductors, by which it may be led away to do useful work.

We therefore fasten to the spindle by which the wire loop is supported a cylinder of metal split into two halves, as shown in Fig. 51, each half being insulated from the spindle. Then we cut the loop at c (Fig. 50) and attach each end of the wire to one half of the cylinder. Finally, we arrange two stationary conducting blocks b b so as to bear on the cylinder and convey the current away to an external circuit.

The split cylinder is termed a **commutator**, and the stationary blocks are termed **brushes**. As the com-

FIG. 51.—Commutator.

mutator turns with the spindle, the slots pass the brushes at the same time as the sides of the wire loop reach the top and bottom of the dotted circle. The result is that each time the direction of current in the loop reverses, the connexions of the loop to the brushes reverse also, so that the current in the external circuit is always in the same direction.

Since the effect of the induced current is to oppose the motion which produces it, mechanical energy is expended in moving the conductors through the field. This mechanical energy is transformed into electrical energy, part of which is used in forcing the current round the wire loop (where it appears as heat), and part of which is available in the external circuit. The resistance of the wire loop is comparable to the internal resistance of a cell, and causes a similar difference between the e.m.f. generated and the potential difference across the external circuit.

Motor Effect

The effect we have been considering is reversible, and instead of using the motion across a field to produce current, we can use a current to produce motion across a field.

Consider the conductor shown in cross section in Fig. 52 (left). It is situated in a magnetic field, the direction of which we will assume to be from left to right. If the conductor is carrying current, it is surrounded by a field of its own, which would normally be represented by the dotted circles. We will suppose that the current is flowing down into the paper, so that the direction of this field is clockwise.

In the space above the conductor (nearer the top of the page) the two fields assist each other, while in the space below it they oppose. The result is to produce a combined field similar to that shown on the right of the figure.

It is convenient, in considering the action of magnetic fields, to think of the lines as tending to straighten out

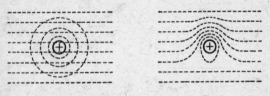

Fig. 52.—Conductor in Magnetic Field.

like stretched elastic threads, while at the same time trying to get as far away from each other as possible. The result in the present case is to cause the conductor to move towards the bottom of the page. The force exerted on the conductor is proportional to the flux density and the current.

Fig. 53 shows in diagrammatic fashion the direction of motion for two directions of current. It should be compared with Fig. 49, when it will be seen that the direction of motion is always the opposite of that which would have produced the current.

Accordingly, if we use the left hand instead of the right, and let the First finger point in the direction of the Flux and the seCond finger in the direction of the Current, the thuMb will indicate the direction of Motion. This is the " left-hand rule ".

Note that if either the direction of the current or the direction of the field is reversed, the direction of motion is reversed also, but that if the directions of both the current and the field are reversed, the direction of motion remains the same.

Part of the electrical energy expended is used in forcing the current through the wire, where it appears as heat. The remainder is available for forcing the wire against any mechanical resistance which impedes its motion, that is, in doing mechanical work.

A machine for converting electrical to mechanical energy is termed a **motor**. An attempt might be made to run the elementary generator we have already described as a motor by connecting a supply of current to

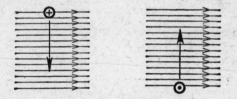

FIG. 53.—Motion of Conductor.

the brushes. Since, however, both brushes are likely to make momentary contact with the same half of the commutator when the slots pass beneath them, there would be a serious danger of short-circuiting the supply. Moreover, if the motor happened to stop in this position, it would not start again.

We shall see in the next chapter how these difficulties can be overcome.

QUESTIONS

1. What conditions are necessary for the generation of current by the motion of a conductor across a magnetic field ?

2. What is meant by the term *flux-linkage* ?

3. What general law governs the direction of an induced current ?

4. What flux density is required to produce an e.m.f. of 1 volt in a conductor 25 centimetres long moving at

right angles to a magnetic field at a speed of 800 centimetres per second ?

5. Explain how the " right-hand rule " enables the
direction of an induced current to be found.

6. What is the object of fitting a commutator to a
generator ?

7. What happens when current is passed through a
conductor lying at right angles to a magnetic field ?

8. Explain why the internal resistance of (a) a generator
and (b) a motor causes part of the applied energy to be
wasted.

9. Explain how the " left-hand rule " enables the direction of motion of a conductor in a magnetic field to be
found.

CHAPTER IX

D.C. GENERATORS AND MOTORS

WE must now see how the principles outlined in the preceding chapter can be applied to practical generators and
motors. For the present we shall confine our attention
to generators in which the current supplied to the external
circuit flows always in the same direction, and to motors
suitable for operation on such current. Machines of
this kind are called **direct current (d.c.)** generators and
motors. Sometimes the term continuous current is used
instead of direct current, and sometimes generators are
called dynamos.

Practical Generators

The output of a single wire loop rotating in a magnetic
field is very small. It is also far from uniform, for
although the two-section commutator ensures that the
current in the external circuit flows in only one direction,
it does not alter the fact that during each half revolution
the current grows gradually from zero to a maximum,
and then dies away again.

In an attempt to produce a steadier current, we might
arrange two wire loops at right angles, so that instead of
two conductors cutting the field we had four, as shown
in cross-section in Fig. 54. One pair of conductors
would then produce their maximum e.m.f. while the other

pair were producing their minimum, and if we could arrange for the two pairs to be connected in series, we should have a current which, while still not uniform, was much steadier than before.

FIG. 54.—Rotation of Conductors.

In practice, this principle is carried farther, and a large number of conductors are arranged on the periphery of a drum and rotated together. Fig. 55 shows eight such conductors (four loops) in cross section. The

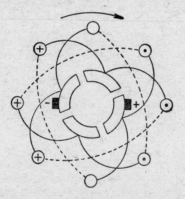

FIG. 55.—Conductors and Commutator.

distant ends of the loops are shown by dotted lines and the connexions of the near ends to the commutator by solid lines. The commutator has four sections instead of two. The brushes are represented by solid blocks.

Suppose that the rotation is clockwise, and that the direction of the field is again from left to right. Then all the conductors on the left-hand side will carry current flowing down into the paper, and all those on the right-hand side current flowing up out of the paper.

The action can be better understood if we imagine each pair of conductors to be replaced by an electric cell. Since the conductors are sources of e.m.f., this is a permissible assumption. If the connexions of Fig. 55 are traced out, they will then be found equivalent to those in Fig. 56. From this it is apparent that between one brush and the other there are two parallel paths, in each of which are two sources of e.m.f. in series.

Note that if the external circuit is not closed, the e.m.f. of one branch opposes that of the other.

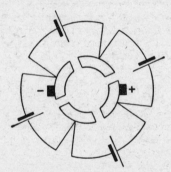

Fig. 56.—Equivalent Circuit.

Each of the four cells in Fig. 56 represents two conductors in Fig. 55. As the eight conductors are evenly spaced round the circular path, the total e.m.f. is reasonably steady, and as new conductors are continually taking up the positions shown, the current in the external circuit is always in the same direction.

We have supposed so far that the conductors are surrounded by air. In order to obtain a magnetic path of low reluctance, it is desirable that as much as possible of the space not occupied by conductors should be filled with iron. The conductors are therefore placed in slots on a cylindrical iron core, the whole being termed an armature. Fig. 57 is a cross section of an armature core having twelve slots. Except in the smaller sizes, internal channels are provided for ventilation.

Rectangular coils of several turns may be used instead of single loops, all the turns of a coil being accommodated in the same pair of slots. As the number of slots and the number of commutator sections are often large, the connexions of a complete armature may appear very complicated, but the principle is similar to that of the simple example we have described.

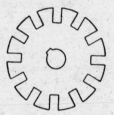

FIG. 57.—Armature Stamping.

Commutators are usually built up from copper sections (segments) clamped together on the shaft and insulated from each other and from the shaft by mica. Brushes are made from special grades of carbon, and are allowed to bed themselves down on to the curved surface of the commutator. The brushes slide in brush-holders, and are kept in contact with the commutator by means of springs.

Eddy Currents

As the iron armature is rotated with the conductors, it too cuts the magnetic field, and if proper precautions were not taken, considerable currents would be induced in it. These **eddy currents** would circulate in the iron, heating it up and wasting energy.

The armature is therefore laminated, or made up of thin sheets or stampings. The stampings are of the general shape shown in Fig. 57, a stack of them being clamped together and mounted on the armature shaft.

Each stamping has a thin coating of insulation on one side (sometimes the natural coating of oxide is sufficient), so that eddy currents are confined to individual sheets and the building up of an appreciable e.m.f. along the length of the armature is prevented.

Eddy currents can be further reduced by using iron of high ohmic resistance. Alloys having this property were mentioned on page 80.

Field Magnet

The magnet which produces the field in which the conductors rotate is called the **field magnet**. It is nearly always an electromagnet, and in order to effect a further

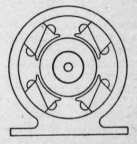

FIG. 58.—Polepieces and   FIG. 59.—Four-Pole
Armature.   Machine.

improvement in the magnetic circuit it is fitted with iron **pole-pieces,** shaped so as to come as close to the armature as possible (Fig. 58).

Although the minimum number of poles on the field magnet is naturally two, one north and one south, it is usually desirable to have more than two, so that each conductor passes north and south poles alternately several times during one revolution. The number of poles is an important factor in the design of the generator. A four-pole field magnet with the armature in position is shown in outline in Fig. 59.

Armature Reaction

We have assumed so far that the only magnetic field present is that of the field magnet. Actually, the

armature has a field of its own while current is flowing
in its winding, and this field is at right angles to that of
the field magnet.

The result is to distort the original field so that it
appears to have moved round slightly in the direction
in which the armature rotates. Thus, in Fig. 60, which
represents an armature between two pole-pieces, f is
the direction of the field due to the field magnet, a the
direction of the field due to the armature, and c the
combined effect of the two.

The effect of the armature upon the field is known
as the **armature reaction**. As the correct position of
the brushes depends upon the direction of the field, it is
necessary to move them round too; if this were not
done, serious sparking would occur. The amount by
which they are moved is called the **angle of lead**.

For reasons which will be mentioned in Chapter X,
however, this is not always enough to ensure sparkless
commutation, particularly when
the current drawn by the external
circuit varies from time to time.
It is therefore a common practice
to provide the field magnet with
interpoles. These are small auxiliary poles placed between the
main poles and having windings
arranged in series with those of
the armature. As their strength

Fig. 60.—Armature
Reaction.

depends upon the armature current, they are able to
provide the necessary compensation.

Field Connexions

The windings of the field magnet are sometimes fed
from a separate source of power, when they are said to
be **separately excited**. The normal practice, however,
is for the machine to supply its own field current. It
is able to do this because the residual magnetism of
the iron allows a small e.m.f. to be generated without
any current in the field-magnet winding. The current
which this e.m.f. causes to flow assists the residual
magnetism, and so the flux and the e.m.f. build up to

their full values. Generators of this kind are said to be **self-excited**.

Fig. 61 shows two ways of connecting the armature *a* and the field-magnet winding *f*. On the left the two are in parallel, and the machine is said to be **shunt wound**. (Almost any parallel connexion is commonly called a shunt.) On the right, they are in series, and the machine is said to be **series wound**. In both cases the small circles represent the generator terminals, *i.e.*, the connexions to the external circuit.

In the shunt-wound generator the voltage across the field winding is the same as that at the terminals. If extra current is taken by the external circuit, the voltage drop in the armature is increased and the terminal voltage decreases. The consequent drop in the field

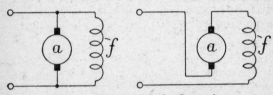

Fig. 61.—Shunt and Series Connexions.

current causes a further fall in the terminal voltage The effect is particularly marked at heavy loads.

In the series-wound generator, if extra current is taken by the external circuit, extra current flows through the field winding. Owing to the rapid rise in flux for small increases in current (page 77), the terminal voltage, within certain limits, rises as the current taken increases. This renders the series-wound generator unsuitable for most purposes, but it is sometimes used as an auxiliary generator to compensate for the voltage drop in a long cable. It is then called a **booster**.

The voltage of a generator can be regulated by hand if means are provided for altering the strength of the field. Thus in the shunt-wound generator the field strength may be controlled by a variable resistance in series with the field-magnet winding.

It is often required that the terminal voltage of a generator should remain constant on various loads without regulation by hand. Since the terminal voltage of a shunt-wound machine decreases as the load increases, while that of a series-wound machine increases, a combination of the two can be made to give a nearly constant voltage. The field magnet then has two windings, one in series with the armature and one in parallel, and the generator is said to be **compound wound.**

Motors

We mentioned in the preceding chapter that the simple wire loop connected to a two-section commutator was hardly suitable for running as a motor, owing to the supply being short-circuited when both brushes made contact with the same half of the commutator. This difficulty does not arise when the number of commutator segments is increased, and most commercial d.c. generators can be made to run as motors.

The effect of armature reaction in a motor is opposite to that in a generator, and in order to allow for it, the brushes are given an angle of lag instead of an angle of lead; that is, they are moved from the mid position in a direction contrary to that of rotation. As in the case of the generator, however, interpoles are often employed to ensure sparkless commutation.

When the armature of a motor rotates, the conductors cut the field, and an e.m.f. is produced in them just as it is in the case of a generator. This **back e.m.f.** opposes the **applied e.m.f.,** the difference between them being the voltage which drives the current through the armature resistance. If we multiply the back e.m.f. by the armature current we obtain (neglecting losses) the electrical equivalent in watts of the mechanical power developed by the machine.

Motors, like generators, may be either series, shunt or compound wound. The speed of a shunt-wound motor does not vary much with changes in load, but that of a series-wound motor falls as the load increases. A series-wound motor is more suitable for starting under heavy loads, but is liable to " race " if the load is suddenly removed.   Compound-wound motors have characteristics between the other two.

Note that the direction of rotation of a motor is not reversed by reversing the supply current, because this would change the polarity of both the field and the armature.   To reverse the rotation it is necessary to reverse the connexions of either the field or the armature, but not both.

The speed of a motor can be varied by varying the strength of the field, the speed increasing as the field is weakened.   The field strength can be controlled by means of a variable resistance connected in series with the field-magnet winding of a shunt-wound machine or in parallel with that of a series-wound machine.   Alternatively, the speed of either type of machine can be controlled by a resistance in series with the armature, but this leads to heavy losses.

Motor Starters

When a motor is running, the back e.m.f. is nearly as great as the applied e.m.f., but at the moment of starting there is no rotation, and therefore no back e.m.f.   It follows that if the normal voltage were suddenly applied, a greatly excessive current would flow through the armature.

Motors of any size are therefore provided with a starter.   This is a tapped resistance connected to a series of contact studs so arranged that when a switch arm is moved over the studs from an " off " position the resistance is gradually cut out of the armature circuit.   The circuit is closed while the resistance is all in circuit, and the arm is then brought slowly over the studs as the machine starts up.

In order to ensure that the starter is used on every occasion, it is necessary to prevent the switch arm from being left in its final position when the current is switched off.   In the starters commonly used for shunt-wound motors, the arm is held in its final position by a small electromagnet known as a low-voltage release.   The winding of this magnet is included in the field-magnet circuit, and when the supply is switched off (or if for any other reason the field-magnet circuit is broken) it no longer holds the switch arm, which is then returned to its " off " position by a spring.

Very often, a second small electromagnet, known as an overload release, is connected in series with the machine. Should the current increase unduly, the overload release operates contacts which short-circuit the low-voltage release, and again the arm returns to its " off " position.

Losses in Generators and Motors

The object of a generator is to convert mechanical into electrical energy, and that of a motor to convert electrical into mechanical energy. Owing to losses in the machine, however, not all the energy put in is available in the changed form. The following are the chief sources of loss :

Copper losses caused by the passage of current through the resistance of the armature and field-magnet windings. The lost energy appears as heat in the windings.

Iron losses caused by eddy currents (page 93) and hysteresis (page 79). In both cases the lost energy appears as heat in the iron parts of the machine.

Mechanical losses caused by air resistance and bearing friction. In this case, too, the lost energy is ultimately converted into heat.

The smaller these losses, the greater is the *efficiency* of the machine. The statement that a machine has an efficiency of 90% at full load means that, at the load for which it is designed, nine-tenths of the energy put into it is available for external use in the changed form.

Example.—*The combined copper, iron and mechanical losses in a generator total* 1200 *watts. If it supplies* 20 *amperes at* 240 *volts, what is its efficiency?*

$Output = 20$ *amps at* 240 *volts* $= 4800$ *watts.*
$Input = 4800 + 1200 = 6000$ *watts.*
$Efficiency = (4800/6000 \times 100)\% = 80\%.$

Large machines are more efficient than small ones, and the efficiency of a large generator may approach 95%.

QUESTIONS

1. A wire loop rotating in a magnetic field produces a very unsteady current. How can this defect be overcome?

2. What are *eddy currents*? How can they be kept at a minimum in a generator armature?

3. What is meant by *armature reaction*?

4. What is meant by the terms *shunt-wound, series-wound* and *compound-wound* as applied to generators and motors?

5. How do the characteristics of a shunt-wound generator differ from those of a series-wound generator?

6. How do the characteristics of a shunt-wound motor differ from those of a series-wound motor?

7. What is meant by the term *angle of lead* as applied to the brushes of a generator? Compare the positions of the brushes on generators and motors.

8. What is the *back e.m.f.* of a motor?

9. Why is a motor starter necessary?

10. Two 100-volt motors have efficiencies of 72% and 80% respectively. If their mechanical outputs are equal and the first one takes a current of 10 amperes, what current is taken by the other?

11. A motor having an efficiency of 78% drives a generator having an efficiency of 75%. If the input to the motor is 12 amperes at 200 volts, what is the output of the generator in watts?

CHAPTER X

ELECTRICAL INDUCTANCE

Mutual Induction

We have seen that an e.m.f. is induced whenever the magnetic flux linked with an electric circuit is changed, and that one method of producing such a change is to cause a conductor forming part of the circuit to cut lines of force, either by moving the conductor itself or the magnet producing the flux.

It is possible, however, for a change in flux linkage to occur without actual movement. Consider the two single-turn coils in the centre of Fig. 62, one of which can be put in series with an electric cell *c* by means of a key *k* while the other is permanently connected to a galvanometer *g*. We will call the coil connected to the cell the primary coil and the one connected to the galvanometer the secondary. There is no electrical connexion between one coil and the other.

When the key is closed and current flows in the primary coil, the galvanometer gives a momentary deflection, showing that for a short time current flows in the secondary coil also. We can explain this effect by considering the magnetic field set up by the current in the primary.

The lines of force form closed paths round the conductor, as described in Chapter I. Some of them are represented in the present figure by dots, although they do not exist, of course, until the key is closed. It is

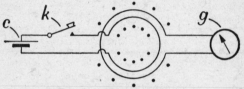

FIG. 62.—Mutual Induction.

clear that when the magnetic field has been set up, there are lines of force linked with the secondary coil which were not there before, and it is this change in flux linkage which produces the momentary current through the galvanometer.

When the key is opened, the current in the primary coil ceases and the magnetic field disappears. Once again there is a change in flux linkage and a momentary current in the secondary, but this time the deflection of the galvanometer is in the opposite direction. Evidently the current in the secondary flows in one direction when the primary circuit is closed and in the opposite direction when it is opened.

This is in accordance with Lenz's law (page 83), which states that induced currents always tend to oppose the change which produces them. Thus, when the primary circuit is closed, the current in the secondary tries to prevent the flux from being set up, but when the primary circuit is opened it tries to prevent it from dying away.

A similar result would be obtained if the outer coil were used as the primary and the inner as the secondary, or if two coils of equal size were placed together. The effect would be more pronounced if each coil consisted

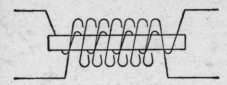

FIG. 63.—Primary and Secondary Coils.

of many turns, particularly if the flux were increased by winding them upon an iron core, as suggested in Fig. 63.

Self-Induction

In dealing with the motor, we said that when the armature conductors were moving across the magnetic field, an e.m.f. was induced in them, just as it would have been had the machine been a generator. We called this the back e.m.f., because it opposed the e.m.f. applied to the motor.

A similar action occurs in the present case, for as the magnetic field builds up, the flux linkage changes not only in the secondary coil, but also in the primary. Let us, therefore, ignore the secondary and consider what happens when a single coil, say a solenoid, is connected to a source of current.

As soon as the current starts to flow, the magnetic field starts to build up, and the flux linked with the coil changes. The result is to produce an e.m.f., called the back e.m.f. of self-induction, which opposes the applied e.m.f. Were it not for the back e.m.f., the current would rise instantaneously to its value as determined by Ohm's law, thus:

$$\text{Current} = \frac{\text{Applied e.m.f.}}{\text{Resistance}}.$$

As it is, however, the back e.m.f. retards the growth of

current, and the rise takes place gradually. In Fig. 64, in which vertical heights represent fractions of the final (Ohm's law) value of current and horizontal distances represent time, the way in which the current rises is shown by curve *r*.

A circuit in which these effects are marked is said to be **inductive**. The rate at which the current starts to grow (that is, the slope at the beginning of the curve) is determined by the applied voltage and by a property of the particular circuit known as its **self-inductance**, or often simply as its inductance. As the current grows, the *rate* of growth becomes less. There is there-

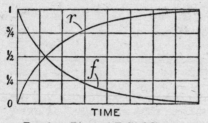

FIG. 64.—Rise and Fall of Current.

fore a reduction in the rate of change of flux linkage and of the back e.m.f. As the back e.m.f. falls, more and more of the applied e.m.f. is available for forcing the increased current against the resistance of the circuit.

Theoretically, the current never quite reaches its Ohm's-law value in an inductive circuit, but it does so for all practical purposes within a comparatively short time. According to the inductance and resistance of the circuit, this time may be anything from a small fraction of a second upwards. An increase in resistance, while reducing the value of the final current, also reduces the effect of inductance in retarding the current growth.

Unit of Inductance

The unit of inductance is the **henry** (plural, henrys; not henries), named after Joseph Henry, the American physicist. A circuit has an inductance of 1 henry if a current which is changing at the rate of 1 ampere per second produces in it a back e.m.f. of 1 volt.

The inductance of a coil is dependent upon the linkage between turns and flux. In a single-turn coil the linkage is equal to the total flux, since all the flux links with the single turn. In a coil with two turns there is twice as much flux and also twice as many turns for it to link with, so that the inductance is four times as great. Provided that every line of force links with every turn, we can say that in general the inductance is proportional to the square of the number of turns. Actually, this condition is not always attained in practice, particularly in the case of a long coil such as a solenoid, and some of the flux links with less than the full number of turns.

Note that if iron is present, the flux varies in the manner shown in the curve on page 77, *i.e.*, it is not proportional to the current flowing. The inductance of an iron-cored coil is therefore dependent in part upon the amount of current flowing through it, that is, upon the state of magnetization of the core.

When one coil acts upon another, as in Fig. 62 or 63, they are said to have mutual inductance. This, too, is measured in henrys. Two coils have a mutual inductance of 1 henry when a current changing at the rate of 1 ampere per second in one of them produces an e.m.f. of 1 volt in the other.

The henry is an inconveniently large unit for many purposes, and a subdivision, the **microhenry**, is commonly employed. One henry equals one million microhenrys.

Decay of Current

Let us return to Fig. 64 and suppose that, after the current has grown to its final value, the coil is suddenly short-circuited. We need not concern ourselves with what happens to the source of current, but we will suppose that it is disconnected at the moment that the short-circuit is applied.

If a resistance without self-inductance is short-circuited, the current in it immediately drops to zero. In the present case, however, as soon as the current starts to fall, the magnetic field starts to collapse, and in so doing induces an e.m.f. in the coil. Since the coil is short-circuited, this e.m.f. can cause a current to flow, and in accordance with Lenz's law this current flows in the direction which tends to maintain the flux, *i.e.*, in the same direction as the original current.

The result is that the current dies away gradually, as indicated by curve *f*. As it dies away, the rate of fall becomes less, but in a short time it has dropped practically to zero. As in the case of the rising current, an increase in resistance reduces the effect of the inductance.

Storage of Energy in Field

While the current is dying away, the electrical energy it represents is being converted into heat in the coil. This energy cannot come from nowhere, and it is clearly not coming from the original source, which we have supposed to be disconnected. Actually, it has been *stored* in the magnetic field. When the circuit was first closed, energy was required to establish the field; that is why the current did not rise instantaneously to its full value. Now, as the field collapses, this energy is restored again to the circuit.

We can compare the effect of inductance with that of mechanical inertia, which is the tendency of a body to go on doing what it is doing already, whether moving or standing still. When a train starts, energy is absorbed in " getting it going ", but once it is running at the required speed it can be kept going by supplying just sufficient energy to make good the frictional losses. When the power of the engine is cut off, the train continues to move until the energy which it acquired at the start has been exhausted.

Suppose that instead of short-circuiting the coil, we simply opened its circuit. It would be like stopping the train by putting an obstacle on the line, for in neither case would there be any opportunity for the harmless dissipation of the stored energy.

In the electrical case, the collapsing field would induce the usual e.m.f., and this would force a current to flow by arcing across the switch contacts. We shall refer to the passage of current by arcing in Chapter XVIII ;

for the present we need note only that damage to switches and dangerously high voltages can be caused by suddenly breaking a highly inductive circuit.

Inductance in Generator and Motor Coils

We referred on page 95 to difficulties which arise in obtaining sparkless commutation. In a multi-coil armature, current is generated by a number of coils connected in series, as in Fig. 55. When the commutator segments to which a coil is connected pass a brush, this current has to be suddenly reversed. In the brief time during which the coil is short-circuited by the brush, therefore, the current must die away in one direction and build up again in the other. Owing to the self-inductance of the coil, these operations would normally require more time than is available, and it is in order to assist the current to reverse by inducing an e.m.f. opposed to the e.m.f. of self-induction that interpoles are fitted.

The use of carbon brushes assists towards the same end, owing to the contact resistance included in series with the coil in which the current is being reversed. Carbon brushes have therefore replaced the copper brushes which were used in early machines.

Non-Inductive Windings

We have spoken particularly of the inductance of solenoids and similar coils, because, owing to the flux from one turn threading others as well, this is the case in which inductance is most marked. Nearly all conductors, however, possess inductance in some degree.

When a strictly non-inductive resistance is required, a conductor may be doubled back on itself. It can

then be wound in a coil as indicated in Fig. 65. Since the current flows round the coil in one direction in the

FIG. 65.—Non-Inductive Winding.

first half of the conductor and in the other direction in the second half, the two magnetic fields cancel out and there is no inductive effect.

The symbol shown in Fig. 66 is sometimes used for a

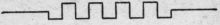

Fig. 66.—Non-Inductive Resistance.

resistance when it is desired to emphasize that it is non-inductive.

Inductances in Series and Parallel

Inductances, like resistances, can be connected in series (Fig. 67) or in parallel (Fig. 68). Provided that

Fig. 67.—Inductances in Series.

they have no mutual inductance, their joint value may be found by the methods used for resistances. These were described in Chapter II.

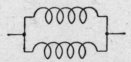

Fig. 68.—Inductances in Parallel.

Example.—*What is the joint inductance of two coils, of 2 henrys and 4 henrys inductance respectively, connected (a) in series and (b) in parallel, there being no mutual inductance between them ?*

In series, joint inductance = (2 + 4) henrys = 6 henrys.

$$In\ parallel,\ joint\ inductance = \frac{1}{0.5 + 0.25}\ henrys$$
$$= 1.33\ henrys.$$

If there is mutual inductance, *i.e.*, if any of the flux of one coil links with any of the turns of the other, the joint inductance may be either more or less than that obtained above. It will be more if the two inductances assist each other (as when the two coils are placed end to end and the current flows in the same direction round each of them), and less if they oppose.

QUESTIONS

1. How can a current be induced in a coil without actual movement?
2. What is mutual induction?
3. What is self-induction?
4. What is inductance, and how is it measured?
5. What happens when an inductive circuit is (a) closed, (b) opened?
6. When a coil carrying current is short-circuited, current continues to flow for a short time. From what source is the energy represented by this current derived?
7. Why is it sometimes dangerous to break a highly inductive circuit suddenly?
8. Explain more fully than was possible in Chapter IX why interpoles are fitted on generators and motors.
9. What is a non-inductive winding?
10. What is the effect of connecting inductances (a) in series and (b) in parallel?

CHAPTER XI

ELECTRICAL CAPACITANCE

WE must now examine another property of electric circuits which, like inductance, comes into prominence when an e.m.f. is applied or removed.

Electrical Capacitance

Suppose that water is flowing along a pipe; obviously the flow can go on indefinitely so long as there is plenty of water and no obstruction in the pipe. This corresponds to the ordinary flow of a steady current in a conductor. If the pipe is obstructed at some point by closing a tap, the flow of water ceases; this corresponds to opening an electric circuit at a switch.

Consider what happens if the pipe is obstructed, not by a rigid tap, but by an elastic diaphragm made, say, of thin rubber. Fig. 69 shows such a diaphragm at d, the pipe being enlarged at the point where it is placed and the whole being represented in section.

Like the rigid obstruction, the diaphragm is a "non-conductor" of water, and there can be no steady flow. When, however, the water pressure which would have caused a flow is first applied, the diaphragm will stretch a little and allow a momentary flow to take place. Thus, if the flow would have been from left to right, the diaphragm will stretch to position a; if from right to left, to position b.

In either case the diaphragm will be strained, and there will be more water on one side of it and less on

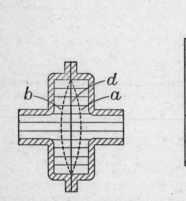

FIG. 69.—Elastic Diaphragm.

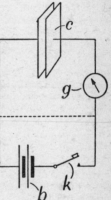

FIG. 70.—Capacitance.

the other. If the piece of the pipe shown in the figure is now removed and the ends sealed, the diaphragm will remain in its strained condition. If the ends are later connected by an open pipe, the diaphragm will spring back to its original position and there will be another momentary flow of water, this time in the opposite direction. The energy stored by virtue of the strained condition of the diaphragm will thus be given up again.

There is an electrical analogy to this action. Let a battery b, Fig. 70, be connected in series with a key k to two flat metal plates c, facing each other but not touching. We will include a galvanometer g in the circuit to show what is happening.

When the key is closed, the galvanometer gives a momentary deflection, showing that although the metallic circuit is open between the plates, current has flowed for a short time. The plates are termed a **capacitor** (formerly a **condenser**), and the capacitor is now said to be **charged**.

Opening the key is like removing the piece of water-pipe and sealing its ends; a movement of electricity has taken place, but the way of retreat has been cut off, so the capacitor remains charged. If a conductor is now connected across the capacitor and galvanometer, as shown by the dotted line, the galvanometer will give a momentary deflection in the opposite direction. There has again been a flow of current, and the capacitor is now discharged.

The space between the metal plates is the electrical equivalent of the rubber diaphragm; although it is a non-conductor of electricity, it can be subjected to an electrical strain which allows a momentary flow of current in the circuit. The substance occupying this space (air, in our case) is called the **dielectric**. All insulators will act as dielectrics, but some, as we shall see, are better than others.

We can regard the capacitor as a device which receives electricity when it is charged and stores it until it is discharged. In the water circuit the quantity of water displaced depends partly upon the pressure applied and partly upon the mechanical characteristics of the diaphragm. Similarly, in the electric circuit, the quantity of electricity stored depends partly upon the voltage applied and partly upon the electrical characteristics of the capacitor. The quantity of electricity stored for a given difference of potential between the plates is a measure of their **capacitance**, so that we may say,

$$\text{Capacitance} = \frac{\text{Quantity of electricity}}{\text{Difference of potential}}.$$

The term **capacity** was formerly used for capacitance.

Unit of Capacitance

The unit of capacitance is the **farad**, named after

Michael Faraday, the discoverer of electromagnetic induction. A capacitor has a capacitance of 1 farad if a charge of 1 coulomb (page 20) corresponds to a difference of potential between one plate and the other of 1 volt. We may therefore write the above equation in the form:

$$\text{Farads} = \frac{\text{Coulombs}}{\text{Volts}}.$$

It follows from this that

$$\text{Coulombs} = \text{Farads} \times \text{Volts},$$

and that

$$\text{Volts} = \frac{\text{Coulombs}}{\text{Farads}}.$$

Like the henry, the farad is an inconveniently large unit for most purposes, and a subdivision, the **microfarad**, is therefore largely used. One million microfarads equal 1 farad.

Permittivity

In the water circuit the characteristics of the diaphragm which affect the amount of water displaced are its area, its thickness, and the material of which it is made. If we look upon these factors as determining the " capacity " of the diaphragm, it will help us to remember that the capacitance of an electrical capacitor is:

    (*a*) proportional to the area of the dielectric under strain,

    (*b*) inversely proportional to the thickness of the dielectric, and

    (*c*) proportional to a quality of the dielectric known as its **permittivity**.

Note that the most important part of a capacitor is the dielectric. While it is under strain, an **electric field** is established in it; this field can be compared with the magnetic fields with which we are already familiar. Like them, it has a definite direction, which can be represented by lines of force (this time, electric, not magnetic), and like them, it has a definite strength at any point. In the present case we may consider its strength to be uniform and its direction straight across the dielectric from one plate to the other (Fig. 71).

The permittivity is a measure of the effectiveness of the dielectric material as compared with a vacuum. If we say that a particular material has a relative permittivity of 4, we mean that a capacitor in which this material is used as a dielectric has four times the capacitance of a similar capacitor in which the dielectric is a vacuum.

Fig. 71.—Electric Field.   Fig. 72.—Multi-Plate Capacitor.

The following are the approximate relative permittivities of some materials commonly used:

Air . . . . practically unity
Paper . . . . 1·5
Insulating oil . . 2
Ebonite . . . 2·5
Mica 5

If A is the area in square metres of the dielectric under strain and d its thickness in millimetres, then for an air dielectric,

$$\text{Capacitance} = 0{\cdot}0088 \; \frac{A}{d} \text{ microfarads.}$$

For other dielectrics the capacitance is increased in accordance with the relative permittivity.

Note that an increase in the thickness of the dielectric decreases the capacitance.

Example.—*What is the value of a capacitor with air dielectric having two plates each 10 centimetres square, the plates being 2 millimetres apart?*

Area of dielectric under strain = area of one plate = 100 square centimetres = 0·01 square metres.

Relative permittivity (of air) = 1.
Thickness of dielectric = 2 mm.

$$Capacitance = 0.0088 \frac{A}{d} \times 1$$

$$= 0.0088 \times \frac{0.01}{2}$$

$$= 0.000044 \text{ microfarad.}$$

In order to increase the area, it is usual to employ more than two plates, alternate plates being connected in sets as shown in Fig. 72. We must remember, however, that it is the area of the dielectric, not the area of the plates, that matters. In the example shown there are four plates on one side, three on the other, and six sheets of dielectric. The capacitance is therefore six times that of a single pair of plates of the same size and spacing.

In order to avoid joining a large number of separate plates, capacitors for telephone and similar purposes

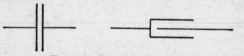

FIG. 73.—Capacitor Symbols.

are often made of long sheets of metal foil (or metallized paper) separated by plain paper, the whole being rolled together and impregnated with paraffin wax.

The symbols shown in Fig. 73 are used in circuit diagrams to represent capacitors.

Energy Stored in Capacitor

The energy required to charge a capacitor is dependent upon the quantity of electricity stored and the potential difference established between the plates. It will be remembered that energy is measured in joules and that

Energy (joules) = Potential Difference (volts)
          × Quantity (coulombs).

In the present case, however, the potential difference and the quantity of electricity have to rise from zero to their full values. Because of this the energy is equal to only half that given by the above expression, and we may write:

Energy (joules) = $\frac{1}{2}$ Potential Difference (volts)
    × Quantity (coulombs).

But we have seen that

Quantity (coulombs) = Capacitance (farads)
    × Potential Difference (volts),

so that

Energy (joules) = $\frac{1}{2}$ Potential Difference (volts)
    × Capacitance (farads)
    × Potential Difference (volts),

$$= \frac{\text{Capacitance (farads)} \times \text{Voltage}^2}{2}.$$

We may look upon this energy as being stored in the electric field set up in the dielectric. We have already found that energy can be stored in a magnetic field, and it is instructive to compare the two cases.

Dielectric Strength

If the pressure in the water circuit were too great, the diaphragm would burst. Similarly, if the electrical pressure applied to a capacitor is too high, the dielectric will break down. The breakdown usually takes the form of a puncture at the weakest point. If the dielectric is air, no permanent harm is done, but a solid dielectric, which would originally have a much higher dielectric strength, will be left with a small hole, which will reduce its future dielectric strength to that of air.

The breakdown voltage increases with the dielectric thickness, but not proportionally. Dielectric strength is measured by the voltage required to break down a sheet 1 millimetre thick, and the following are some typical figures:

Air	4,000 volts.
Ebonite . . .	40,000 volts.
Mica	60,000 volts.

Dielectric strength must not be confused with the

permittivity of the dielectric referred to earlier, nor with the ordinary ohmic resistance (insulation resistance) of the dielectric. Since there are no perfect insulators, there will always be a slight normal flow of current through a solid or liquid dielectric; the equivalent condition in the water circuit would be a slight porosity of the diaphragm. As an appreciable flow of current through the dielectric would allow the charge to leak away, it is important that the insulation resistance should be as high as possible.

Charge and Discharge

Think again of the water circuit, and suppose that nothing is forcing the water either in one direction or the other. Clearly, there will be no difference of pressure between the two sides of the diaphragm. When pressure is applied to the water and the diaphragm is stretched, the pressure across the diaphragm builds up until, when everything is steady again, it is equal to the applied pressure. Before this can happen, however, some water has to flow, and the pressure across the diaphragm therefore builds up gradually. The time it takes depends upon the amount of water which has to flow and the resistance which it meets on the way.

Similarly, in the electric circuit in Fig. 70 there is no potential difference between the plates until the key is first closed. The potential difference (provided that the circuit offers some resistance) then builds up gradually until the capacitor is fully charged, when it is equal to the applied e.m.f. The current which flows at any moment is dependent upon the difference between the applied e.m.f. and the capacitor e.m.f., so that the current gradually falls as the capacitor is charged. This condition should be compared with the gradual *rise* in current and *fall* in back e.m.f. in an inductive circuit.

When a capacitor is discharged through a resistance, the current again starts at a maximum, and dies away as the energy stored in the capacitor is converted into heat in the resistance. At the same time the voltage falls, so that when the capacitor is fully discharged it is again zero.

Capacitance in Cables

A cable consists of one or more conductors in an insulating covering. If there are two separate conductors, they act as the two plates of a capacitor, the insulation between them forming the dielectric. The cable therefore possesses capacitance. A similar effect occurs in the case of a single conductor using an earth return, particularly if the cable is enclosed in an earthed metal sheath or is laid underground or in water.

The effect of the capacitance is to delay the growth of current at the other end until the cable itself has been charged. This delaying action is of special importance in telegraph and telephone cables, in which signalling is carried on by intermittent or rapidly varying currents. There is an appreciable capacitance even between bare overhead conductors such as those used in high-voltage power transmission lines.

Electrolytic Capacitors

We saw in Chapter IV that one of the troubles of primary cells is polarization, *i.e.*, the formation on one of the electrodes of a thin insulating film which tends to prevent current from flowing. A similar action is turned to good account in the electrolytic capacitor. In its simplest form this consists of two aluminium plates immersed in an appropriate electrolyte. When an e.m.f. is applied, an insulating oxide film appears on one plate, and the device then acts as a capacitor.

As the film is very thin, electrolytic capacitors can provide a large capacitance in a small space, but they are not always suitable for high voltages, or for cases in which the applied e.m.f. may be in either direction. Commercial forms are of a semi-dry type, often containing absorbent material impregnated with the electrolyte.

Capacitors in Series and Parallel

Capacitors may be connected in series (Fig. 74) or in parallel (Fig. 75). When they are in parallel, they are equivalent to a single capacitor with larger plates, and the total capacitance is found by adding the individual capacitances together.

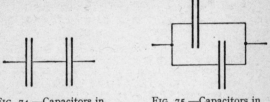

FIG. 74.—Capacitors in Series   FIG. 75.—Capacitors in Parallel.

Example.—*What is the joint capacitance of a 1-microfarad capacitor connected in parallel with a 10-microfarad capacitor?*

Capacitance = (1 + 10) *microfarads* = 11 *microfarads*.

When capacitors are connected in series, they are equivalent to a single capacitor with thicker dielectric. If they are of equal capacitance, the joint capacitance is found by dividing the capacitance of one of them by the number of capacitors. If they are not equal, the procedure is the same as for resistances in *parallel*.

Example.—*What is the joint capacitance of an 8-microfarad capacitor in series with a 4-microfarad capacitor?*

$$Capacitance = \frac{1}{0 \cdot 125 + 0 \cdot 25} \ microfarads$$
$$= 2 \cdot 67 \ microfarads.$$

Note that the calculation for capacitors in parallel is like that for resistances in series, and vice versa.

QUESTIONS

1. What is meant by electrical capacitance? In what piece of apparatus is it prominent?

2. When a capacitor is discharged, energy is given up. From what source is this energy derived?

3. What is the unit of capacitance and what does it represent?

4. What is the relation between the capacitance of a capacitor, the voltage applied to it, and the quantity of electricity stored in it?

5. How do the dimensions of a capacitor affect its capacitance?

6. What is meant by *permittivity*?

7. Find the approximate capacitance of a capacitor the dielectric of which is a single sheet of mica 3 centimetres long, 2 centimetres wide and 0·1 millimetre thick.

8. Upon what does the energy stored in a capacitor depend?

9. How does the current vary when a capacitor in series with a resistance is (a) charged and (b) discharged?

10. What is meant by dielectric strength? How does it differ from insulation resistance?

11. What is an electrolytic capacitor?

12. What is the joint capacitance (a) in series and (b) in parallel of two capacitors, one of 0·5 microfarad and the other of 0·25 microfarad capacitance?

13. What is the joint capacitance of ten 1-microfarad capacitors connected in series?

14. A capacitor has a capacitance of 0·5 microfarad. What will its capacitance be if the thickness of the dielectric is doubled?

CHAPTER XII

ALTERNATING CURRENTS

WE saw in Chapter VIII that the current generated in a wire loop rotating in a magnetic field flowed first in one direction and then in the other. In order to prevent similar reversals in the current flowing in the external circuit, we made connexion with the wire loop through a two-section commutator.

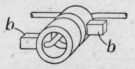

FIG. 76.—Slip Rings.

If we had not been concerned about the nature of the current in the external circuit, we could have used a pair of insulated **slip rings**, such as those shown in Fig. 76, instead of the commutator, and arranged for one of the brushes *b b* to bear on each ring. The current in the external circuit would then have kept on reversing in direction in the same way as that in the wire loop.

Current of this kind is termed **alternating current** or **a.c.** It is simpler to generate and transmit than direct current, and for most purposes is just as useful. Since, however, it must necessarily be always dying away in one direction and then building up again in the other, the effects of inductance and capacitance described in the last two chapters are of much greater importance than in the case of direct current.

Generation of Alternating Current

In order to fix our ideas, let us consider the rotation in a magnetic field of the single conductor *c* shown in the left-hand part of Fig. 77. We can, if we like,

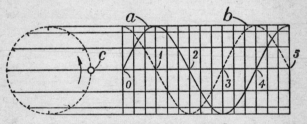

FIG. 77.—Generation of Alternating Current.

imagine that this conductor forms one side of a wire loop; it will not affect the argument. The path in which the conductor moves is shown by the dotted circle, and the direction of rotation (counter-clockwise) by the arrow. The direction of the field is supposed to be from left to right, so that (as we can verify by applying the right-hand rule) the direction of current is *into* the paper.

Since the generation of current depends upon the cutting of lines of force, the part of the circular movement in which we are most interested is the motion *across* the field, *i.e.*, the " up-and-down " motion that would be seen by a spectator away on the right-hand side of the page. Let such a spectator plot a curve showing how the position of the conductor changes from time to time as it moves across the field.

The curve that he will obtain is shown at *a* on the right of the figure. Horizontal distances represent *time*, and vertical distances the position of the conductor on either side of the horizontal centre line from which it starts. Let us suppose that one complete revolution takes 4 seconds, so that the horizontal divisions marked 0, 1, 2, etc., represent seconds. In the diagram, vertical " time lines " are drawn for each third of a second.

Starting with the conductor in the position shown, the movement is upward for the first quarter of a revolution, the highest point being reached at the 1-second line. Intermediate points on the curve are found by drawing horizontal lines from the position of the conductor in the circle to the corresponding time line.

For the second quarter of the revolution the movement is downward, so that at the 2-second line the conductor is back again at the centre line. For the next quarter the downward movement continues, until at the 3-second line the lowest position is reached. For the fourth quarter the movement is upward, and at the 4-second line the conductor is once more on the centre line. In the figure the curve is continued for a further second, this part representing the beginning of the next revolution.

The e.m.f. generated in the conductor is dependent upon the rate of cutting the lines of force, that is, upon the speed at which it moves across the field. The *distance* that it moves across the field is represented by vertical heights on the curve, and since time is represented by horizontal distances, the *rate* at which the lines are being cut at any moment is indicated by the slope of the curve.

Thus, at the start, the conductor is moving straight across the field, and the rate of cutting is a maximum. The curve is therefore steep. As the conductor moves round the first quarter of the circle, the rate diminishes until the top is reached, when there is no cutting of the field, and for a moment the curve is horizontal. Then, during the next quarter of the circle, the rate of cutting again increases to a maximum, this time in the opposite direction, as indicated by the increasing *downward* steepness of the curve. The rate then diminishes, and the curve becomes less steep until it is horizontal for a moment as the conductor passes the bottom of the circle. Finally, the rate of cutting increases to a maximum again in the first direction as the conductor returns to its original position.

The curve *a* which we have been considering is known to mathematicians as a sine curve. One of the properties of a sine curve is that its *steepness* from point to point is represented by the *height* of an exactly similar curve displaced horizontally by one quarter of a revolution. The dotted line *b* is such a curve. It crosses the centre line when curve *a* is momentarily horizontal, *i.e.*, when the rate of cutting lines of force is zero. It reaches its greatest distance from the centre line when curve *a* is sloping most steeply, *i.e.*, when the rate of cutting lines of force is a maximum. Moreover, its position above or below the centre line is an indication of whether curve *a* is sloping upwards or downwards, that is, of the direction in which the conductor is moving across the field.

Since curve *b* gives a complete picture of the rate and direction of motion of the conductor across the field, it also represents the magnitude from moment to moment of the induced e.m.f.

Commercial a.c. generators, like commercial d.c. generators, use many conductors instead of one. Nevertheless, the e.m.f. which they produce is usually not very different from that represented by a sine curve, and as this curve lends itself to fairly simple mathematical treatment, it forms the basis of most calculations.

Frequency

In the above example, curve *b*, like curve *a*, goes through a complete cycle of changes during each revolution of the conductor. The number of cycles which occur in a second is known as the **frequency** of the current. One cycle per second is called one **hertz** (abbreviated Hz).

Although, for the sake of simplicity, we assumed that one revolution took 4 seconds, an alternating current which took such a long time to complete one cycle would be of little value, and in practice the length of a cycle is only a small fraction of a second. The frequency of most alternating-current power supplies in this country is fifty cycles per second or 50 Hz.

Effective Value of Current

We cannot expect a current that is continually dying away and building up again to be as effective as a steady current equal to its maximum strength. Evidently we need some sort of average value of the alternating current before we can find its direct-current equivalent.

There are different kinds of average. Consider the four squares in Fig. 78, and suppose that the lengths

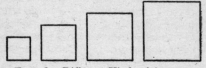

FIG. 78.—Different Kinds of Average.

of the sides are 2, 3, 4 and 5 metres respectively. The average length of side is then

$$\frac{2 + 3 + 4 + 5}{4} \text{ metres} = 3 \cdot 5 \text{ metres.}$$

The areas of the squares are 2 × 2, 3 × 3, 4 × 4 and 5 × 5 square metres, or 4, 9, 16 and 25 square metres, respectively. The average area is therefore

$$\frac{4 + 9 + 16 + 25}{4} \text{ square metres} = 13 \cdot 5 \text{ square metres.}$$

A square of this area has a side which measured in metres equals the square root of 13·5, and this is not 3·5, but about 3·67. We have therefore found two kinds of average: one the simple average of the sides of the squares, and the other the side of a square equal to the average area. Both are equally correct, and either might be of greater interest in any particular case. If we were putting fences round square fields, for instance, we should be concerned with the first kind of average, but if we were buying turf with which to cover them, we should be concerned with the second.

We have investigated this point because the effective value of a current is proportional to the *square* of its strength. Thus, on page 63, we saw that the heating effect of a current is not twice but four times that of half the current flowing through the same resistance.

Now, to find the average value of, say, one half-cycle of alternating current (Fig. 79), we might draw a large number of vertical lines, add them together, and divide the result by the number of lines. If the rise and fall of current followed a sine curve, the result would be 0·637 of the maximum value.

FIG. 79.—Finding Average Current.

Clearly, however, this is not the sort of average we want. To find the effective value, we must square (multiply by itself) the length of each vertical line, divide the total by the number of lines to find the average square, and then find the square root of the result to obtain the effective current. For the sine curve, this sort of average is not 0·637 but 0·707 of the maximum value. It is called the **root-mean-square (r.m.s.)** value, because it is found by taking the square root of the mean (average) of all the squares. Another name for it is the **virtual value**.

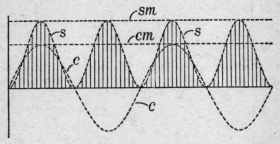

FIG. 80.—Root-Mean-Square Value.

Fig. 80 will make this important but rather difficult point clearer. Curve *c* is a sine curve representing two cycles of alternating current. Curve *s* is obtained by taking a number of points on curve *c*, squaring their distances above or below the centre line, and plotting the results to any convenient vertical scale. Since heights on curve c above the centre line represent current flowing in one direction, while depths below it represent current flowing in the other, it is permissible to consider the former as positive values and the latter as negative. Note that curve s lies wholly above the centre line; this is because the squares of both positive and negative quantities are positive.

Since heights on curve s represent the squares of the current values, the shaded area represents the sum of these squares. It is evidently equal to half the area under the line sm, which represents the square of a steady current cm equal to the maximum value of the alternating current. We may say, therefore, that

$$\text{Effective value of current}^2 = \frac{\text{Maximum value of current}^2}{2}.$$

To come back to curve c, we take the square root of these quantities and find that

$$\text{Effective value of current} = \frac{\text{Maximum value of current}}{\text{Square root of } 2}.$$

To find the effective value of an alternating current, we therefore divide the maximum value by the square root of 2, or 1·414. This is the same as multiplying it by 0·707, the figure already given. The same rule applies in the case of an alternating e.m.f.

Example.—*What is the effective value of an alternating current which varies in accordance with a sine curve and has a maximum strength of fifteen amperes?*

Effective value = 0·707 × maximum value
= 0·707 × 15 amps. = 10·6 amps.

Wave Form

Although it is convenient for purposes of calculation to assume that the rise and fall of an alternating current can be represented by a sine curve, this is not always the case in practice. The actual shape of the curve is known as the **wave form**. Fig. 81 shows two possible

wave forms, one more flat-topped than a sine curve
and one more peaked.

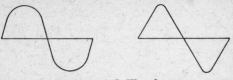

FIG. 81.—A.C. Waveforms.

Electrical Angles

When, as in the simple case we have been considering,
one revolution of the conductor represents one cycle of
e.m.f., it is natural to divide the cycle into 360 degrees.
In practical generators of alternating current one
revolution usually represents more than one cycle, but
the practice of dividing the *cycle* into 360 degrees is so
convenient for mathematical purposes that it is still
followed. As a reminder that the angles measured by
these degrees do not necessarily correspond to the angles
actually turned through by the rotating part of the
generator, the former are sometimes spoken of as **electrical
angles** and the degrees as **electrical degrees.**

QUESTIONS

1. Explain how alternating current may be generated by
the movement of a conductor in a magnetic field.
2. How does the rate of cutting lines of force vary as
the conductor in Question 1 rotates?
3. What is a sine curve?
4. What is meant by the frequency of an alternating
current? What is the standard frequency in this country?
5. Why does the effective value of an alternating current
differ from the average value?
6. What is meant by the abbreviation r.m.s.?
7. What is the maximum value of an alternating e.m.f.
the r.m.s. value of which is 230 volts?
8. What is meant by the division of a cycle of alternating
current into electrical degrees?
9. Define the terms (*a*) frequency, (*b*) wave-form.
10. What is the r.m.s. value of an alternating current
having a maximum value of 7 amperes?
11. How many times per second does current having a
frequency of 50 Hz rise to a maximum?

CHAPTER XIII

THE A.C. CIRCUIT

Effect of Resistance

LET us see what happens when alternating current is passed through a resistance. The circuit will be as shown in Fig. 82, in which the symbol on the left represents the source of current, say an a.c. generator.

To find the current at any moment, we divide the voltage by the resistance in accordance with Ohm's law.

As the resistance remains the same, the current is proportional to the voltage, and as the latter is continually changing in strength and direction, the current changes in a similar manner.

The e.m.f. and current curves therefore rise and fall together, as shown at e and c in Fig. 83. It does not matter which curve rises to the greater height; this depends upon the vertical scales we choose to employ for volts and amperes. What does matter is that both curves are of the same frequency, and that when one

FIG. 82.—A.C. and Resistance.

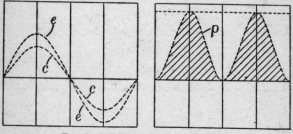

FIG. 83.—Effect of Pure Resistance.

of them reaches a maximum or minimum, the other does the same. When the e.m.f. and current reach their maximum and minimum values at the same time in this manner, they are said to be **in phase**, which simple means in step, with each other.

The power (watts) in the circuit at any moment is found as usual by multiplying volts by amperes. If we do this at different points along the curves and plot the results, we shall obtain a curve such as that shown at p on the right of the figure. This curve relates to the same period of time as the curves on the left, but is shown separately to avoid confusion.

It is clear that the power reaches a maximum at the same time as the voltage and current. Note, however, that the power curve lies wholly above the zero line; we can explain this mathematically by saying that whether volts and amperes are both negative or both positive, their product is positive. We had a similar case in the last chapter when we were finding the effective value of an alternating current.

In Fig. 83, the total energy (power × time) expended during the cycle is represented by the shaded area, and as this is half the total area under the horizontal dotted line, the average power is half the maximum.

Effect of Inductance

Now suppose that the source of alternating current is connected to an inductive circuit (Fig. 84), the resistance of which is small enough to be neglected. We saw in Chapter X that when a current changes in an inductive

Fig. 84.—A.C. and Inductance.

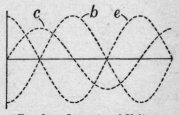

Fig. 85.—Current and Voltage Curves.

circuit an e.m.f. is induced, and it will be remembered that the value of this e.m.f. at any instant is dependent upon the rate of change of current and that its direction is such as to oppose the change.

Let the current be represented by curve c in

Fig. 85. The steepness of this curve at any point indicates the rate at which the current is changing, and can be represented by a similar curve displaced horizontally by 90° or one quarter of a cycle. Such a curve is shown at *b*, and this represents the induced e.m.f.

The tendency of the induced e.m.f. is to prevent the current from changing, and as the current does change, the applied e.m.f. must be opposing the induced e.m.f. at every instant. In the present case this is all the applied e.m.f. has to do, since we have assumed that the resistance of the circuit is negligible. The applied e.m.f. can therefore be represented by curve *e*, drawn equal and opposite to the induced e.m.f. *b*.

Note that in this circuit we have the curious condition of a current reaching its maximum when the

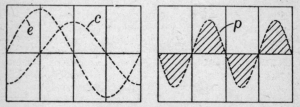

FIG. 86.—Effect of Inductance.

voltage which causes it to flow is at a minimum, and vice versa. The current is always one quarter of a cycle behind the e.m.f., and is said to **lag** by this amount. Another way of expressing the same thing is to say that the voltage and current are 90° **out of phase**. Do not be misled because at first sight the current curve *c* appears to be ahead of the voltage curve *e*; the current lags because it reaches its maximum or minimum value after the voltage, and in the diagram this means that the current curve is displaced towards the right.

The applied e.m.f. *e* and current *c* are shown again on the left of Fig. 86. In this figure the curves start at the moment when the e.m.f. is zero, but the phase relationship (*i.e.*, the relative positions of the two curves) is just the same as in Fig. 85. Let us see how the lagging current affects the power (watts) in the circuit.

To obtain the power curve p, we again multiply volts by amperes at different points on the curves and plot the results. Evidently the power will be zero when *either* the e.m.f. *or* the current is zero, and as they are out of phase, this occurs four times during each cycle. The power curve for the interval of time we are considering is shown on the right of Fig. 86.

The power curve does not now lie wholly above the centre line, as it did when the circuit was non-inductive. This is because there are times when the e.m.f. is positive and the current negative, and vice versa. The mathematical result of multiplying a positive by a negative quantity is negative, and at such times the power curve is below the line. The energy is again represented by shading, and it will be observed that the areas below the line are equal to those above.

The shaded areas above the line represent energy being taken from the source to build up the magnetic field; note that they coincide with the times when the current is growing. The shaded areas below the line represent energy being restored to the source during the collapse of the field; note that they coincide with the times when the current is falling. We have already met this storage of energy in the magnetic field in Chapter X.

As the positive parts of the power curve are equal to the negative parts, the average value is zero. It follows that although there are both current and e.m.f., no power (watts) is being expended in the circuit. Current which flows under these conditions is sometimes said to be **wattless**.

Effect of Capacitance

The effect of capacitance in an alternating-current circuit can be most readily appreciated by considering again the flexible diaphragm (Fig. 69, page 109) in a water circuit. Clearly, such a diaphragm would not prevent the flow of an alternating water current; it would simply stretch first one way and then the other as the pressure changed. Note, however, that the maximum flow of water would occur when the diaphragm was midway between one stretched position and the other, while the pressure would be at its maximum when the stretched positions were actually reached.

Now suppose that a source of alternating current (Fig. 87) is connected to a capacitor, the resistance of

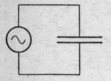

FIG. 87.—A.C. and Capacitance.

the circuit being negligible. The current flowing at any instant (amperes, or coulombs per second) is evidently equal to the rate at which the charge on the capacitor (coulombs) is changing, whether the charge is increasing or decreasing. As the charge is proportional to the voltage (Chapter XI), the current is dependent upon the rate of change of voltage.

Fig. 88, in which curve *e* once more represents e.m.f. and curve *c* current, illustrates these conditions. Starting on the left, the e.m.f. rises from zero to a maximum. The capacitor is then fully charged and the current momentarily zero. As the e.m.f. falls, the capacitor discharges, the current reaching a maximum when the e.m.f. is zero. The current continues to flow in this

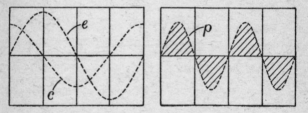

FIG. 88.—Effect of Capacitance.

direction as the e.m.f. builds up in its new direction. It falls as the capacitor charges, until when the e.m.f. has reached its maximum in the new direction, the capacitor is again fully charged and the current has dropped to zero. These operations continue indefinitely.

As in the case of the purely inductive circuit, the current and voltage are 90° out of phase, but this time the current is leading. The power curve for the same interval of time is shown on the right as before, and it will be observed that the average power is again zero. This is explained by the fact that the energy taken from the source to charge the capacitor is given up again when it discharges. In the meantime it has been stored, as we saw in Chapter XI, in the electric field set up in the capacitor dielectric.

Resistance with Capacitance or Inductance

If a resistance is connected in series with the capacitor in the last example, the result is somewhere between the case of pure resistance (Fig. 83) and pure capacitance (Fig. 88). Consequently, the current leads the voltage by some amount less than 90°.

This condition is shown in Fig. 89. Note that the power curve p now encloses a greater area above the

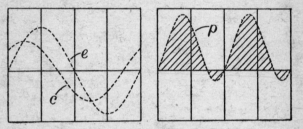

FIG. 89.—Effect of Capacitance and Resistance.

centre line than below it, so that the average power is no longer zero. This is to be expected, since some of the energy is being converted into heat in the resistance.

Similarly, if the circuit includes inductance and resistance, the result is somewhere between the case of pure resistance (Fig. 83) and pure inductance (Fig. 86). Consequently, the current lags behind the voltage by some amount less than 90°, and again the average power is more than zero. It is left to the reader to draw the curves for this case.

Power Factor

Since in an alternating-current circuit it is possible to have current and e.m.f. without any power, it is clearly not possible to find the power in watts simply by multiplying volts by amperes. To find the true power, the product of volts and amperes must be multiplied by a figure which takes the phase difference into account. This figure is known as the **power factor**. It varies from zero in a circuit comprising pure inductance or pure capacitance to unity in a circuit comprising pure resistance. We may say, therefore, that

Watts = Volts × Amperes × Power Factor.

The abbreviation **p.f.** is used for power factor.

Example.—*What is the true power in a circuit connected to 230-volt a.c. mains, the current being 10 amperes and the power factor 0·85?*

Power = (230 × 10 × 0·85) *watts* = 1955 *watts*.

Those who are familiar with elementary trigonometry may note that in all the cases we have been considering the power factor is equal to the cosine of the angle representing the phase difference between current and voltage.

The product of volts and amperes (neglecting the power factor) is called the **apparent power** and measured in volt-amperes, or kilovolt-amperes for large loads, in order to distinguish it from the **true power** measured in watts or kilowatts. The abbreviation kVA is used for kilovolt-amperes.

Reactance

Inductance and capacitance, like resistance, impose a limit on the current which a given voltage will cause to flow in a circuit, but their effect, unlike that of resistance, varies according to the frequency of the current. The frequency of a steady direct current is zero, and to it an inductance acts as an ordinary conductor and a capacitance as a complete break. As the frequency increases, the inductance offers a less and less easy path, and the capacitance a more and more easy one.

For current of any given frequency, the effect of either an inductance or capacitance may be compared with that of a resistance and expressed in ohms. This value is termed the reactance of the inductance or capacitance at that frequency. It follows that if there is no real resistance in the circuit,

$$\text{Current} = \frac{\text{Voltage}}{\text{Reactance}}.$$

Evidently the reactance of an inductance increases as the frequency increases, while that of a capacitance diminishes. The actual relations are, for an inductance,

Reactance (ohms)
$$= 2\pi \times \text{Frequency} \times \text{Inductance (henrys)},$$

and for a capacitance,

Reactance (ohms)
$$= \frac{1}{2\pi \times \text{Frequency} \times \text{Capacitance (farads)}}.$$

The Greek letter π (pi) represents the ratio between the circumference and diameter of a circle, its value being rather more than 3·14. It comes into the equations because the reactance is dependent upon the number of radians through which the conductor generating the current moves in one second. A radian is the angle obtained by measuring along the circumference of a circle a length equal to its radius, and one revolution is therefore equivalent to 2π radians.

Impedance

Since real circuits usually possess appreciable resistance in addition to their inductance or capacitance, this must be combined with the reactance in order to find the total equivalent resistance at any given frequency. The total equivalent resistance is called the **impedance** and, like the reactance, can be expressed in ohms. We may therefore write:

$$\text{Current} = \frac{\text{Voltage}}{\text{Impedance}}.$$

The impedance is not the simple arithmetical sum of the reactance and resistance, but the *square* of the impedance can be found by adding the square of the reactance to the square of the resistance, from which it follows that

$$\text{Impedance} = \sqrt{\text{Reactance}^2 + \text{Resistance}^2}.$$

The ratio of resistance to impedance is another way of expressing the power factor, so that

$$\text{Power Factor} = \frac{\text{Resistance}}{\text{Impedance}}.$$

It is obvious from this relation that in a circuit with no resistance the power factor is zero, while in one with nothing but resistance (impedance = resistance) it is unity. This is in accordance with the results already obtained.

Resonance

A consideration of the combined effects of inductance and capacitance is beyond the scope of the present book, but we may note that they can in some cases counteract each other. For example, it is possible to have a combination of inductance and capacitance in series in which the current is in phase with the total applied voltage. The reactance is then zero, and the current, being limited only by the resistance of the circuit, is a maximum. For any given combination of inductance and capacitance there is a definite frequency at which this will occur, and the circuit is said to be **resonant at** that frequency.

QUESTIONS

1. What are the relations between current and voltage in an alternating-current circuit comprising :
>   (a) resistance only,
>   (b) inductance only,
>   (c) capacitance only,
>   (d) resistance and inductance,
>   (e) resistance and capacitance ?

2. Upon what does the power in an alternating-current circuit depend ?

3. What is meant by the term *power factor* ?

4. In a certain 250-volt circuit the current is 10 amperes and the power is two kilowatts. What is the power factor ?

5. What do you understand by (a) reactance and (b) impedance ?

6. The resistance of a circuit is 300 ohms and its reactance at a frequency of 50 Hz is 400 ohms. What is its impedance at the same frequency?

7. What is the power factor when current having a frequency of 50 Hz flows in the circuit in Question 6?

CHAPTER XIV

A.C. GENERATORS AND MOTORS

A.C. Generators

The outline of a d.c. machine shown in Fig. 59 (page 94) will serve also for a small alternating-current generator, or **alternator**. Alternators, of course, have no commutator, and the direct current for energizing the field-magnet must be obtained from a separate source. A small d.c. generator termed an exciter is sometimes mounted on the same shaft for this purpose.

As in the d.c. generator, the field-magnet poles, if there are more than two, are alternately north and south, and since a complete cycle of current is generated every time a conductor passes a north and south pole in succession, the frequency of the current is given by the equation:

Frequency = Revolutions per second × Number of pairs of poles.

The armature carries several groups of conductors, the groups being spaced around the periphery so as to occupy similar positions in relation to the field-magnet poles. The current is collected from the armature by means of slip-rings.

Since all that is necessary for the generation of current is relative motion between conductors and field, a generator could be made to work by holding the armature and rotating the field-magnet round it. This would be very inconvenient, but a similar effect can be obtained by placing the field-magnet system on the central rotating part, and the conductors in which the current is generated on the surrounding stationary part. As the field-magnet windings then rotate, they must be connected to the d.c. source by a pair of slip-rings.

Part of an alternator in which the conductors move and the field-magnets are stationary is shown on the left of Fig. 90, while the corresponding arrangement, in which the field-magnets move and the conductors are stationary, is shown on the right. In both cases the slots in which the conductors are housed are indicated, but the conductors themselves are omitted for the sake of clearness.

In order to avoid confusion, the terms **stator** and **rotor** are used instead of armature and field-magnet. The stator is the stationary part and the rotor is the rotating part, no matter which of them carries the armature conductors and which the field-magnet system.

The advantage of the arrangement shown on the right of Fig. 90 is that the slip-rings and moving windings have to deal only with the comparatively low voltage and small current necessary to produce the field. As the

FIG. 90.—Types of Alternator.

conductors in which the alternating current is generated are stationary, their insulation is simplified, while the fact that connexion can be made to them without slip-rings and brushes removes another difficulty in the generation of high voltages. Alternators of this type are therefore normal and can be made for much larger outputs than are practicable in d.c. generators.

Note that the right-hand rule (page 85) for finding out in which direction an induced current flows assumes that the conductor is moving across the field, and not vice versa. When the conductor is stationary and the field is moving, the rule must be applied as though the direction of motion were reversed.

Polyphase Currents

From the generating point of view, it is better to produce a steady current than one that is continually varying. In the d.c. generator we were able to do this by spacing a number of conductors round the armature, so that while some were generating their maximum e.m.f., others were generating their minimum. Something of the same kind can be done in the case of an alternator, by causing it to generate at the same time two or more currents *differing in phase*. The different currents must be taken from the machine over different circuits, and we are thus led to a **polyphase** system comprising several more or less independent supplies of the same frequency, as distinct from a **single-phase** system having only one supply.

Polyphase systems also enable the winding space on the generator to be utilized more efficiently. Moreover, they simplify the design of alternating-current motors, and lead to economies in the amount of copper required for transmission lines.

Suppose that we provide an alternator of the kind illustrated on the right of Fig. 90 with two sets of con-

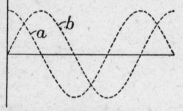

Fig. 91.—Two-Phase Currents.

ductors, the grouping in relation to the spacing of the rotating field-magnets being such that one set is producing its maximum voltage when the other is producing its minimum. We then have two independent sources of current, the phase relationship of the voltages being represented by curves *a* and *b* in Fig. 91. This is a two-phase system, and the difference in phase is one quarter of a cycle, or ninety electrical degrees.

In a similar manner, we can provide a generator with three independent groups of conductors, thus obtaining three separate sources of current, the phase relationship of the three voltages being as shown at *a*, *b* and *c* in Fig. 92. This is a three-phase system, and the difference in phase is one third of a cycle, or 120 electrical degrees.

Instead of fitting a three-phase generator with six output terminals, we can connect one end of each winding to a common point, thus using only four. The arrangement will then be as shown on the left of

Fig. 93, in which the coils represent the three generator windings, and the centre point the fourth terminal. Conductors *a*, *b* and *c* are connected to the terminals at the free ends of the windings. The common return path to the centre point is shown dotted. Conductors *a*, *b* and *c* are known as the three **lines**, and the common centre point (which is usually earthed) as the **neutral point**.

Suppose that three exactly similar loads are connected, one between each of lines *a*, *b*, *c* and the dotted conductor. The three currents will then be equal, and their phase relationship will be the same as that of the voltages. They can therefore be represented by the curves in Fig. 92, from which it will be seen that when

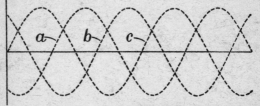

FIG. 92.—Three-Phase Currents

any one current is at a maximum, the other two are half-way towards a maximum in the opposite direction, and that when any one is zero, the other two are equal and opposite. Similar conditions apply at all points on the curves, so that the sum of the three currents at any moment, taking their directions into account, is zero.

It follows that so long as the loads remain the same, there is no current flowing in either direction in the dotted conductor, which under these conditions could be omitted. Each of the three lines is then acting in turn as a return path for the other two. Even if the loads are not the same, the dotted conductor has to carry only the difference between the current flowing outwards and that flowing inwards over the three lines at any instant. It can therefore be of smaller size than the others.

Instead of connecting each load between one of the lines *a, b, c* and the dotted conductor, we can connect one load between *a* and *b*, one between *b* and *c*, and one between *c* and *a*. The voltage applied to each load is then derived from two of the generator windings, but as the two voltages do not reach their maximum at the same time, the joint value is not twice that of one winding, but some smaller figure. The actual value is $\sqrt{3}$, or 1·732, times the voltage of one winding. The current in each line is obviously equal to the current in one winding.

Generator windings arranged as shown on the left of Fig. 93 are said to be **star connected**. The voltage

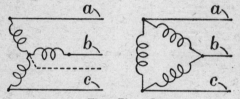

FIG. 93.—Three-Phase Connexions.

between any two of the lines *a, b* and *c* is called the **line voltage,** and that between any one of them and the neutral point the **phase voltage.**

Example.—*The line voltage in an alternating-current system supplied by a star-connected generator is* 400. *What is the phase voltage ?*

$$Line\ voltage\ = Phase\ voltage \times 1·732 = 400.$$

$$Phase\ voltage = \frac{400}{1·732} = 230.$$

An alternative arrangement of the generator windings is shown on the right of Fig. 93, and in this case they are said to be **delta connected** or **mesh connected**. The term " delta " is taken from the Greek capital letter Δ. There is no tendency for current to circulate round the closed path because the sum of the voltages at any instant is zero. The line voltage is that produced by one winding, and is therefore equal to the phase voltage.

The current in each line, however, is $\sqrt{3}$, or 1·732, times the current in one winding.

Note that the two methods of connexion (star and delta) apply not only to the generator windings, but also to the loads. Thus, in the first case we considered, if the loads are connected between each of lines *a*, *b*, *c* and the dotted conductor, they are star-connected, but if between *a* and *b*, *b* and *c*, and *c* and *a*, they are delta connected. This can readily be seen by drawing an example.

A.C. Motors

A.C. generators, like d.c. generators, can be made to run as motors, but only when the frequency of the supply is in step with the frequency at which the armature conductors pass the pairs of poles. The motor must therefore be rotated by some other means until it is running fast enough to continue in synchronism with the supply frequency.

Machines designed to operate in this manner are called **synchronous** motors. Motors which do not operate in synchronism with the supply frequency are said to be **asynchronous** (*i.e.*, not synchronous). The most important class of asynchronous motors depends upon a special property of polyphase currents which we shall now examine.

Rotating Fields

Consider the two pairs of coils shown in Fig. 94. If coils *a* are energized, there will be a magnetic field in line with their axis; let us call this direction north and south. If coils *b* are energized, there will be a magnetic field in line with *their* axis; let us call this direction east and west.

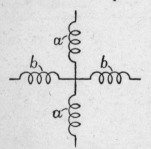

FIG. 94.—Two-Phase Windings.

Suppose now that the coils are connected to a two-phase supply. We can use Fig. 91 to represent the two currents, each curve corresponding to the similarly lettered coils.   Starting on the left-hand side of Fig. 91, current a is at a maximum in one direction; let us assume that this produces a flux in coils a towards the north.   At this moment, current b is at zero, so there is no flux in coils b.   The flux at the centre may therefore be represented by the arrow in sketch 1 of Fig. 95.

A quarter of a cycle later, current a is at zero and current b at a maximum; let us assume that the latter produces a flux in coils b towards the west.   This flux is represented by the arrow in sketch 3 of Fig. 95.

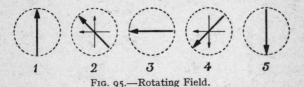

FIG. 95.—Rotating Field.

Half-way between these positions, curve a is still some distance from zero, and curve b some distance from its maximum.   This is the point at which the curves cross, and the two currents are therefore equal.   The two small arrows in sketch 2 of Fig. 95 represent these conditions, and their combined effect is to produce a field towards the north-west as shown by the heavy arrow.

This combination of two fields should be noted.   It follows from the fact that the lines of force cannot have more than one direction at the same place and time. An analogy may be helpful.   Suppose that we set up two electric fans at right angles, so that one produces a wind towards the north and the other a wind towards the west.   If only the first fan is blowing, a particle caught in the wind will move north.   If only the second fan is blowing, it will move west.   If both fans are blowing, its tendency will be to move north-west.

As the field from coils a dies away and that from coils b grows, the combined field, starting from sketch 1, passes through all the intermediate positions to sketch 2, and then through all the intermediate positions to sketch 3.   We have, therefore, a rotating field produced by fixed coils.   Moreover, it can be shown mathematically that the strength of the field does not vary, being always equal to that produced by one of the coils when the current in the other is at zero.

It is important to note that the rotating field is not a matter of approximation, or of a sudden jump from north to west, or even from north to north-west. The rotation of the field is quite smooth and regular, owing to the gradual dying away of flux in one direction and its equally gradual building up at right angles.

Continuing the sequence, the current in coil *b* dies away again after reaching its maximum, and the field towards the west gradually weakens. At the same time, coil *a* is energized by a growing current in the direction opposite to that which produced a field towards the north. This produces a field towards the south, which, in conjunction with the weakening field towards the west, produces a field passing through the south-west as shown in sketch 4 of Fig. 95. When the current in coils *a* has reached a maximum in this direction and that in coils *b* is at zero, the field is towards the south, as shown in sketch 5.

FIG. 96.—Three-Phase Windings.

We have now traced the changes for a complete half-cycle. If the process is continued for the next half-cycle, it will be found that the rotation continues through south-east, east, and north-east until the field is again towards the north. The number of revolutions per second made by the field is therefore equal to the frequency of the current.

We have examined the production of a rotating field by two-phase current because this is the easiest case to follow without detailed mathematical statement. A similar effect can, however, be produced by three-phase currents. In this case three sets of coils are needed as shown in Fig. 96, and the field rotates through 120° between the maximum in one set of coils and the maximum in the next. As this time represents one third of a cycle, the field again rotates at the supply frequency.

Induction Motors

The fact that a rotating field can be produced by stationary coils has led to the development of the induction motor. In this machine, coils carried by the stator produce the rotating field and the rotor is provided with heavy copper conductors accommodated in the usual slots. In many cases these conductors have no external connexions, but are simply joined together at each end of the rotor by heavy copper rings. The term **squirrel-cage** is often applied to rotors of this type.

As the magnetic field rotates, it cuts the rotor conductors and induces currents in the circuits completed by the copper end rings. These currents produce a field which reacts with the rotating field. In accordance with Lenz's law, the effect is to produce motion which will tend to prevent the change of flux linkage, *i.e.*, to make the conductors follow the field.

The rotor therefore turns. It does not, however, actually reach the speed of the rotating field; if it did there would no longer be any induced currents, and there would be no power even to overcome the friction of the motor bearings. The difference between the speed of the rotor and that of the field is known as the **slip**, and at full load may amount in practice to, say, 4% of the speed.

The squirrel-cage rotor without slip-rings is very simple and robust, but is not suitable for starting under heavy loads. Some induction motors are therefore fitted with wound rotors. The winding is in three sections, and is brought out to three slip-rings. The object of the slip-rings is to enable starting resistances to be included in the rotor circuit. These resistances are gradually cut out as the motor speeds up, until finally the slip-rings are short-circuited. The running conditions are then similar to those of the squirrel-cage machine.

Miscellaneous A.C. Machines

In addition to the machines we have described, the following types may be mentioned.

Single-phase Induction Motors.—Although a single-phase supply cannot produce a rotating magnetic field in the way described for polyphase currents, it is nevertheless possible to design a single-phase motor operating on the induction principle. Machines of this kind are not very efficient, but in small sizes they have come into increasing use in recent years. They are not naturally self-starting, but can be made so at the cost of some complication.

Magneto Generators.—These are miniature generators in which the field is produced by permanent magnets. The armature (rotor) is usually of the simple two-slot form shown in Fig. 97, and is known as an H-armature.

FIG. 97.—Magneto Armature.

One cycle of alternating current is generated during each revolution. Magneto generators have been used for the generation of ringing current in some telephone systems, and for ignition current in internal-combustion engines. In the latter case provision is made for interrupting the armature current periodically. The result is to induce a very high voltage in a second winding, also carried by the armature. The high-voltage current so made available is led to the sparking-plugs, where it produces the spark which ignites the explosive mixture in the cylinders.

Commutator Motors.—Since in a d.c. motor, reversal of both field and armature connexions at the same time does not alter the direction of rotation, it is possible to run such motors on alternating current. Owing to the greater tendency to eddy-current losses, however, all the iron, including the field magnet, should be laminated. Small machines of this kind suitable for either d.c. or a.c. supplies are made and are called **universal** motors.

Miniature Synchronous Motors.—For driving electric clocks, miniature synchronous motors are used. They must be operated from mains on which the frequency is "controlled," *i.e.*, kept at or near its stated value over long periods. The motor then runs at a known speed, so that by means of reduction gearing it can be made to operate the hands of a clock. Some but not all of these motors are self-starting.

Miniature synchronous motors of slightly larger size are commonly used for rotating gramophone turntables.

QUESTIONS

1. Upon what does the frequency of the current generated by an alternator depend?

2. What is meant by the terms *rotor* and *stator*?

3. In an alternator, what advantages are gained by making the field-magnet system rotate and keeping the conductors in which the alternating current is generated stationary?

4. What are polyphase currents, and what are their advantages?

5. Draw curves to represent the voltages in (*a*) a two-phase and (*b*) a three-phase system.

6. In what ways can the windings of a three-phase generator be arranged in relation to the lines?

7. The phase voltage of a three-phase star-connected generator is 200. What is the line voltage?

8. What is a synchronous motor?

9. Explain exactly how a rotating field can be produced by stationary windings.

10. What is an induction motor?

11. Why are induction motors sometimes fitted with slip rings?

12. Give an example of an alternating-current generator in which the field is produced by a permanent magnet.

CHAPTER XV

TRANSFORMERS, CONVERTERS AND RECTIFIERS

WE shall deal in this chapter with devices for changing the voltage or nature (a.c. or d.c.) of a supply of electricity.

Advantages of High Voltage

Assuming for simplicity a power factor of unity, we have volts × amperes = watts. If power of, say,

100 kilowatts is transmitted at 10,000 volts, the current is 10 amperes; if at 250 volts, it is 400 amperes. The larger the current, the lower the resistance of the line conductors must be if undue heating losses are to be avoided. Low-resistance lines require a large amount of copper, and copper is expensive.

High-voltage lines have problems of their own (insulation, for example), but their cost is much less than that of a low-voltage line capable of transmitting the same amount of power. From an economic point of view, therefore, it is important that when power is to be transmitted from one place to another, a high voltage should be employed.

Very high voltages are difficult to generate directly, and still more difficult to use safely. A means of converting current generated at one voltage to a higher voltage for transmission, and then reconverting it to a lower voltage for distribution to consumers, is therefore of great commercial importance.

One of the advantages of alternating current is that it can be readily "transformed" from one voltage to another without the use of rotating machines. It is therefore a relatively simple matter to step the voltage up for transmission and then down again for distribution. Voltages up to 132,000 or more are used for transmission in this country, but the voltage of, say, a domestic consumer's supply may be only 240.

Transformers

The device by which the voltage of alternating current is changed is called a **transformer**. We saw in Chapter X that when the strength of the current flowing in a coil alters, an e.m.f. is induced in any other coil the turns of which are linked to the changing flux. Suppose that an alternating current flows in the first coil, say the inner coil of the pair shown in Fig. 63 (page 102). As the current rises, an e.m.f. is induced in one direction in the outer coil, and as it falls, an e.m.f. is induced in the opposite direction. An alternating current in the inner coil therefore produces an alternating e.m.f. of the same frequency in the outer coil, the relation between the two corresponding to that between curves *c* and *b* in Fig. 85 (page 127). If the circuit of the outer coil is completed, the induced e.m.f. will cause a current to flow.

The winding connected to the source of alternating current is called the primary winding, and that in which the e.m.f. is induced, the secondary winding.

The magnetic circuit in Fig. 63 is not very good, as the lines of force in the core have to complete their closed paths through the air. Moreover, there is likely to be a fair amount of magnetic leakage, all the flux produced by the primary not being linked with all the turns of the secondary. In practical transformers a closed iron path is employed. Fig. 98 shows sections

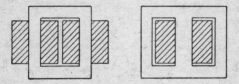

FIG. 98.—Transformers.

of two types, the shaded areas indicating the space occupied by the windings. In both cases the core is laminated, a number of thin stampings being stacked together to provide the necessary cross-sectional area.

The form shown on the left is known as a **core-type transformer.** Half of each winding is placed on each limb, the primary and secondary being either wound one on top of the other, or else split into sections arranged alternately on the core. The form shown on the right is known as a **shell-type transformer.** The core has three limbs, the centre one of which carries the windings, while the outer two form parallel return paths for the magnetic flux.

We will now examine the relations between the voltage and current in the primary and secondary windings. For the sake of simplicity, we will assume first of all that:

(*a*) all the flux produced by the primary winding threads all the turns of the secondary;

(b) the resistance of the windings is negligible in comparison with their reactance at the supply frequency;

(c) the secondary circuit is open.

Since the rate of change of flux is the same for a primary turn as for a secondary turn, the ratio of the secondary e.m.f. to the back e.m.f. of self-induction in the primary (page 102) must be equal to the ratio of the number of turns on the secondary to the number of turns on the primary. As, however, we have assumed the resistance to be negligible, the back e.m.f. is equal to the applied e.m.f., so that we may say:

$$\frac{\text{E.m.f. induced in secondary}}{\text{E.m.f. applied to primary}} = \frac{\text{Number of turns on secondary}}{\text{Number of turns on primary}}.$$

In other words, the primary and secondary voltages are proportional to the numbers of turns. The ratio of secondary to primary turns is known as the **transformation ratio**.

Now suppose that the secondary circuit is closed through a resistance, so that current is drawn from the transformer. Still neglecting losses, and the original small primary current, we may write:

Primary current × Primary turns
    = Secondary Current × Secondary turns

from which it follows that:

$$\frac{\text{Secondary current}}{\text{Primary current}} = \frac{\text{Number of turns on primary}}{\text{Number of turns on secondary}}.$$

In practice there are losses due to the resistance of the windings (proportional, it will be remembered, to the square of the current), and other losses due to eddy currents and hysteresis. The eddy-current loss, but not the hysteresis loss, is largely avoided by the use of a laminated core. In addition, there may be an appreciable amount of magnetic leakage, some of the lines of force produced by the primary taking short cuts and thus failing to link with every turn of the secondary.

None of these losses need be very high, and the efficiency of a well-designed transformer, as measured by the ratio of power output to power input, may be 95% or more in the larger sizes. As, however, the losses that do occur appear as heat, either in the windings or in the core, some provision is necessary to prevent an undue rise of temperature. In very small transformers it is sufficient to employ a well-ventilated case, but larger ones are immersed in insulating oil. Besides carrying the heat away, the oil improves the insulation of the windings and prevents the ingress of moisture.

When a transformer is required to maintain a constant secondary voltage with varying secondary currents, the **regulation** becomes a matter of importance. This is the change in the secondary terminal voltage between

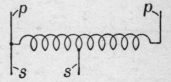

Fig. 99.—Transformer Symbol.

Fig. 100.—Auto-Transformer.

no load (open secondary circuit) and full load, and is usually given as a percentage. We have, therefore:

Percentage regulation =
$$100 \; \frac{\text{Difference in secondary volts between no load and full load.}}{\text{Secondary volts at no load}}$$

The percentage regulation depends in part upon the power factor of the load.

The symbol shown in Fig. 99 is commonly used to represent a transformer in circuit diagrams.

Three-phase currents may be transformed by means of a separate transformer on each phase or by means of a special three-phase transformer. The latter has a separate limb of the core and a separate primary and secondary winding for each phase. As in the case of generator windings, the three primaries and three secondaries may be either star or delta connected.

Auto-Transformers

It is possible to combine the primary and secondary windings of a transformer as shown in Fig. 100, in which all the turns of a single winding are included in, say, the primary circuit p p, while some only are in the secondary circuit s s. A device of this kind is known as an **auto-transformer**.

Auto-transformers can be used for obtaining a small increase or decrease in voltage, but are not suitable when a large transformation ratio is required. In any case, their employment is often prohibited by the fact that they provide a direct metallic connexion between the primary and secondary circuits.

Motor Generators

The conversion of direct current from one voltage to another is not largely practised, but is necessary in some cases. One method is to cause a motor suited to the supply voltage to drive a generator capable of producing the voltage required. A combination of this kind is called a **motor generator**.

The two machines are sometimes built as a single unit, each with its own armature and field-magnet system. As, however, both machines are subject to copper and iron losses (page 99), their combined efficiency compares unfavourably with that of the transformer used in similar circumstances for alternating current.

For the conversion of direct to alternating current at the same or a different voltage, a d.c. motor may be coupled to an a.c. generator. In a similar manner, alternating current may be converted to direct current by coupling an a.c. motor to a d.c. generator.

Rotary Converters

These are machines for the conversion of a.c. to d.c. (or sometimes d.c. to a.c.) in which there is only one field-magnet system and one armature with a single winding. The armature is fitted with a commutator at one end and with slip-rings at the other.

In an ordinary direct-current motor the armature currents flow first in one direction and then in the other, and the conditions are similar when a rotary converter is connected to a d.c. supply. Alternating current may therefore be drawn off by means of the slip-rings, which are connected to tappings on the armature winding. If, as is more usually the case, it is required to convert alternating to direct current, the machine is run as a synchronous motor by connecting the alternating supply to the slip-rings, direct current then being drawn from the commutator.

Rectification of Alternating Current

Alternating current could clearly be converted to direct current if means were available for suppressing alternate half-cycles, the current curve then being as shown on the left of Fig. 101. If, instead of suppressing alternate half-cycles, we could reverse their direction, the result would be as shown on the right.

FIG. 101.—Rectified Current.

The conversion of alternating to direct current in this manner is known as rectification. If alternate half-cycles are suppressed, it is half-wave rectification; if they are reversed, it is full-wave rectification. The current obtained is admittedly intermittent. particularly in half-wave rectification, but as it is always in the same direction, it can be " smoothed " if necessary by connecting a capacitor across the supply to act as a reservoir to bridge the gaps.

There are several devices which conduct electricity in one direction, while being non-conducting, or nearly so, in the other, and any of these can be used for rectification. One such device is the mercury-arc rectifier, in which the one-way conducting path is through mercury vapour contained in a glass bulb or other chamber. Mercury-arc rectifiers can deal with comparatively large currents, and are used for power work. Special patterns are made for connexion to polyphase supplies.

Small currents, on the other hand, can be rectified by thermionic valves, the action in this case being dependent upon the fact that in an evacuated bulb containing two electrodes one of which is heated, current can pass across the gap in one direction, but not in the other.

Small and medium currents can be dealt with by what are commonly called metal rectifiers. When intimate electrical contact is made between a metallic conductor and certain semi-conductors, notably cuprous oxide and selenium, it is found that current will pass across the junction much more freely in one direction than in the other. Although the voltage which can be dealt with

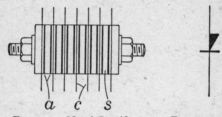

FIG. 102.—Metal Rectifier.　　　FIG. 103.—Rectifier Symbol.

by a single junction is not very large, it is easy to arrange a number of junctions in series.

When cuprous oxide is used, the active element consists of a film of the oxide formed on one side of a copper disc. In the selenium type, a layer of selenium is supported on a metal disc and has its other surface in contact with a disc of soft metal alloy. A number of rectifying elements are assembled as shown in Fig. 102. In addition to the active discs a, there are metallic spacers s and cooling fins c, the whole being clamped together on a threaded metal stem from which the parts are separated by an insulating sleeve.

The symbol shown in Fig. 103 is used in circuit diagrams to represent a half-wave rectifier. The current is supposed to be able to flow out of the point of the triangle, but not into it. This can be remembered by thinking of a funnel, through which water can easily be made to pass in the normal direction, but only with great difficulty in the other.

Four half-wave rectifiers can be connected as shown in Fig. 104 to give full-wave rectification. During one half-cycle current passes through a and b, and during the next through c and d. If the connexions are traced out, it will be found that in each case the current flows in the same direction through any circuit connected across the right-hand pair of conductors. When metal

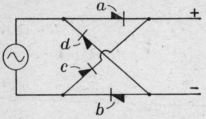

FIG. 104.—Full-Wave Rectification.

rectifiers are used in this manner, the four sets of discs are often mounted on the same centre piece, the general appearance of the whole being similar to Fig. 102.

QUESTIONS

1. Why is it usual to transmit electric power at high voltage?
2. How can the voltage of an alternating current be changed?
3. What is meant by (a) the transformation ratio, and (b) the regulation, of a transformer?
4. In what ways is energy lost in a transformer? What happens to this energy?
5. Explain the action of an auto-transformer.
6. How can the voltage of a direct current be changed?
7. What is (a) a motor-generator, (b) a rotary converter?
8. What are the principles involved in the rectification of alternating current?
9. Give an example of a practical rectifier.
10. Show how metal rectifiers can be arranged to give full-wave rectification.

CHAPTER XVI

ELECTRICAL MEASURING INSTRUMENTS

WE have referred many times to current and voltage, but have not yet seen how these quantities can be measured. It is now time to repair this omission.

An instrument for measuring current (amperes) is called an **ammeter**, and one for measuring voltage a **voltmeter**. Do not confuse the terms voltmeter and voltameter; the latter is the name of a device used in the study of electrolysis. A volt-ammeter, on the other hand, is a combined voltmeter and ammeter.

Connexions of Voltmeters and Ammeters

Most voltmeters really measure current and rely upon Ohm's law for their ability to indicate voltage. Fig. 105

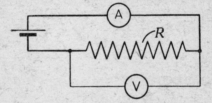

FIG. 105.—Ammeter and Voltmeter.

will make this clear. A cell is sending a current through a resistance R, and it is required to measure the current flowing and the voltage across the resistance. We therefore connect an ammeter A in series with the resistance and a voltmeter V across it.

The ammeter is of very low resistance, and the voltmeter of very high resistance. Apart from this, their construction, which will be described later, may be the same. Each meter has a pointer which is deflected according to the current passing. In the case of the ammeter this current is that flowing in the main circuit, and the meter can therefore be marked in amperes.

As the resistance of the voltmeter is very high, the current passing through it is very small. What little there is depends upon the resistance of the meter and the voltage across it. As the resistance of the meter is a fixed quantity, the current, in accordance with Ohm's law, is proportional to the voltage, which is also the voltage across the resistance R. The meter scale can therefore be marked in volts instead of amperes.

It is important to note that accurate measurements are dependent upon the high resistance of the voltmeter. This is particularly true in the case of a voltmeter connected across a portion only of a circuit; if the resistance of the meter is low enough to allow it to draw an appreciable current, the conditions are the same as those in Fig. 18 (page 37), and the mere connexion of the meter lowers the voltage it is required to measure.

Moving-Iron Instruments

Most ammeters and voltmeters are dependent for their operation upon the magnetic effect of the current. In one type the magnetic field produced by a coil carry-

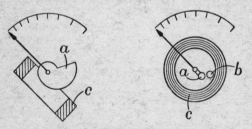

FIG. 106.—Moving-Iron Meters.

ing the current to be measured is made to move a pivoted piece of iron having a pointer attached. Meters of this kind are known as moving-iron instruments.

Fig. 106 illustrates two forms. In that on the left, the coil carrying the current is shown in section at c and the iron is shaped as shown at a. A light spring like the hair-spring of a watch tends to keep the iron and its attached pointer in their normal position.

When current flows in the coil, a magnetic field is set up. The iron tries to move into the densest part of this field, and owing to its shape it rotates about the pivot in doing so, thus moving the pointer over the scale. As the iron moves, the opposition of the control spring increases, until finally the iron and the pointer come to rest. The final position of the pointer depends upon the strength of the field, and therefore upon the strength of the current.

In the form illustrated on the right, the coil carrying the current is again shown at *c*, this time in end view. The moving iron consists of a short rod *a*, only one end of which is seen, attached to the spindle carrying the pointer. As before, a light spring tends to keep the moving parts in their normal position.

A second iron rod *b*, parallel to the first, is fixed inside the coil. When current flows in the winding, similar poles are induced in the adjacent ends of the rods, say two north poles in the ends at which we are looking. The rods consequently repel each other, and the one attached to the pointer moves away from the other until the increasing opposition of the control spring prevents further movement. The pointer moves with it, and indicates the strength of the current.

Moving-iron meters are not very sensitive, and are subject to errors caused by hysteresis. They are, however, simple and robust, and are largely used when great accuracy is not required. Since they operate in the same manner for current flowing in either direction, they are suitable for use on either direct or alternating current. On alternating current the pointer has not time to follow the individual half-cycles of current, and it remains steady in a position corresponding to the root-mean-square value.

The scales of moving-iron instruments are irregular; *i.e.*, the divisions are of different sizes at different points on the scale. This is not always a disadvantage, as it is sometimes possible to arrange for the divisions to be largest on the part of the scale at which the meter is most often used.

Damping

If proper precautions were not taken, the rapid movement of the pointer of an ammeter or voltmeter when current was switched on would cause it to overshoot the mark, and it would then oscillate for a considerable time before settling down in the correct position. As readings would be difficult to take under these conditions, most meters are fitted with some means for damping out the vibrations.

One method is to fit on to the moving system a piston working in a curved box after the manner shown in Fig. 107. The piston does not actually touch the box, but the clearance is small enough to restrict the passage of air from one side to the other. The result is to

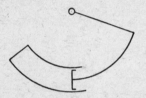

FIG. 107.—Damping Arrangement.

prevent oscillation of the pointer without affecting its final position.

In a method of damping used on some kinds of meter, a light metal disc fitted to the spindle is arranged to move between the poles of a small permanent magnet. As it moves, it cuts the field of the magnet, and currents are induced in it. These currents produce a field which reacts with that of the magnet and, in accordance with Lenz's law, tends to retard the movement. The result is effectively to prevent oscillation, again without interfering with the final position of the pointer.

Meters which are efficiently damped, so that the pointer comes quickly to rest in the correct position, are said to be **dead beat**.

Moving-Coil Instruments

In these meters a coil carrying the current to be measured (or part of it) moves in a magnetic field produced by a fixed permanent magnet. A moving-coil instrument may be compared to a miniature direct-current motor in which the armature never moves more than about a quarter of a revolution.

Consider again the loop of wire in Fig. 50 (page 86), and suppose that it is held in the position shown by means of a light spring. If we pass current round the loop, it will try to rotate, and the distance it will move against the opposition of the spring will depend upon the strength of the current. This is the principle of the moving-coil meter.

Fig. 108 shows the arrangement adopted. The coil *c*, one end of which is shown, consists of a number of turns of fine wire wound on a light metal frame. The field is produced by a permanent magnet of horseshoe form fitted with pole-pieces *N*, *S*. A stationary iron cylinder *i* improves the magnetic circuit and the uni-

FIG. 108.—Moving-
Coil Meter.

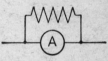

FIG. 109.—Ammeter
Shunt.

formity of the field without adding to the weight of the moving system. The coil is free to rotate between the iron cylinder and the pole-pieces without touching either of them.

Current is led to the coil through two flat, spiral springs (not shown in the drawing), one at each end of the frame on which the coil is wound. These springs serve also to limit the movement of the coil when current is flowing. As the frame on which the coil is wound is made of metal, currents are induced in it as it turns in the magnetic field. It therefore acts in the same manner as the disc mentioned in the last section, and damping is effected without further provision being made.

Moving-coil instruments are accurate and sensitive, and they have an even scale. They are, however, more expensive than those of the moving-iron type. As the direction in which the coil tends to move depends upon the direction of the current, they are not, in their normal form, suitable for use on alternating supplies.

Shunts and Series Resistances

It is often convenient to be able to use the same meter for different ranges of current or voltage. In any case, it is not always possible to wind the meter itself to a resistance suitable for the conditions in which it is to be used. This is especially true of moving-coil meters, in which the available winding space is very limited.

For these reasons, ammeters are often used with shunts, and voltmeters with series resistances, either fixed permanently inside the case or connected externally as required. The effect of a shunt on an ammeter (Fig. 109) is to bypass some of the current, so that more current is needed in the main circuit to produce

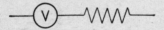

Fig. 110.—Voltmeter Series Resistance.

the same deflection. The effect of a series resistance on a voltmeter (Fig. 110) is to reduce the current which a given voltage will cause to flow, so that a higher voltage is necessary to produce the same deflection.

By way of example, suppose that we have an ammeter that reads up to 1 ampere and we wish to provide a shunt which will make it suitable for use up to 10 amperes. Clearly, nine-tenths of the current will have to flow through the shunt, and only one-tenth through the meter. We therefore make the shunt exactly one-ninth the resistance of the meter and, when using it, multiply all the meter readings by ten.

Or again, suppose that we have a voltmeter that reads up to 10 volts and we wish to provide a series resistance which will enable it to be used up to 100 volts. Clearly, we need a potential difference across the series resistance of 90 volts when that across the meter is 10 volts. We therefore make the series resistance exactly nine times the resistance of the meter, and again multiply all the meter readings by ten.

As we can always increase the range of an ammeter or voltmeter but cannot decrease it, it is an advantage to start off with a sensitive low-range meter. We can, in fact, use such a meter as either a voltmeter or ammeter by fitting suitable resistances and shunts, and this is the principle of the " universal test sets " commonly employed.

Example.—*A moving-coil meter has a resistance of 50 ohms and gives a full-scale deflection* (i.e., *maximum movement of pointer*) *on a current of 1 milliampere. Calculate the necessary resistance of* (a) *a shunt to enable it to be used as an ammeter reading up to 0·1 ampere, and* (b) *a series resistance to enable it to be used as a voltmeter reading up to 100 volts.*

(a) *Current to be measured* $= 0 \cdot 1$ *ampere.*
Current required by meter $= 1$ *milliampere.*
$= 0 \cdot 001$ *ampere.*
Current in shunt $= 0 \cdot 099$ *ampere.*
Resistance of meter $= 50$ *ohms.*
Resistance of shunt $= \frac{50}{99}$ *ohm* $= 0 \cdot 505$ *ohm.*

(b) *Voltage to be measured* $= 100$
Current required by meter $= 0 \cdot 001$ *ampere.*

Total resistance required $= \dfrac{100}{0 \cdot 001}$ *ohms*

$= 100,000$ *ohms.*
Resistance of meter $= 50$ *ohms.*
Series resistance required $= (100,000 - 50)$ *ohms*
$= 99,950$ *ohms.*

Shunts and series resistances should always be made of one of the alloys having a negligible temperature coefficient.

Rectifier and Thermo-Couple Instruments

Moving-coil meters are sometimes fitted with small metal rectifiers to enable them to be used on alternating current. The deflection of a moving-coil meter used in this way does not depend upon the r.m.s. value of the current or voltage, but upon the average value. The difference between the two is explained on page 123. This does not prevent the scale from being marked in the equivalent r.m.s. values, but the readings then need correction if the wave-form is very different from that of a sine curve.

For use on high-frequency alternating currents, thermo-couple instruments are made. These are moving-coil meters arranged to measure the current produced by a thermo-couple (page 67) heated by the current to be measured.

Hot-Wire Ammeters

These are ammeters which measure the strength of a current by its heating effect. Fig. 111 shows the principle of one type. A wire a is stretched between two

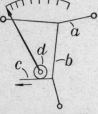

FIG. 111.—Hot-Wire
Ammeter.

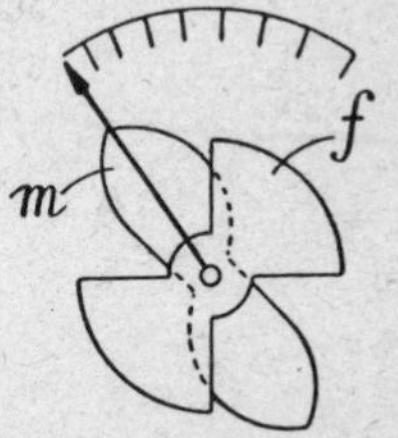

FIG. 112.—Electrostatic
Voltmeter.

fixed points, and a wire b between it and a third fixed point. A silk thread c is attached to the second wire and passed round a pulley d carrying the pointer. A spring tends to pull the thread in the direction of the arrow.

The current to be measured is passed through wire a, which expands as its temperature rises. As it expands, its sag increases, thus allowing the silk thread to increase the sag in wire b. The movement of the silk thread turns the pulley and causes the pointer to move over the scale.

Although hot-wire ammeters are now seldom used, they are of special interest because, being directly dependent upon the heating effect of the current, they read r.m.s. values on a.c. of any frequency or wave form.

Electrostatic Voltmeters

When a capacitor is charged, the electric field produces an attractive force between the plates. This effect is used in the electrostatic voltmeter, the principle of which is shown in Fig. 112. A moving vane *m* can move between a pair of fixed vanes *f*, only one of which can be seen, without touching either of them. The movement is limited by the usual control spring.

The moving vane is connected to one side of the circuit the voltage across which is to be measured, and the fixed vanes to the other side. The combination then acts as a capacitor, and the attractive force between the vanes causes the moving vane to move between the other two to an extent determined by the voltage.

Electrostatic meters are suitable for use on a.c. or d.c. supplies, but not for very small voltages. On a.c. supplies they read r.m.s. values. They take only a very small current on a.c. and none on d.c. except when first switched on.

Wattmeters

The power in a d.c. circuit can be measured by means of a voltmeter and ammeter, but in an a.c. circuit it is necessary to take the power factor into account as well. It is possible, however, to measure power directly by means of a wattmeter. This has a moving coil similar to that of an ordinary moving-coil instrument, but instead of a permanent magnet, another coil is used to produce the magnetic field. One coil is connected in the circuit like a voltmeter, and the other like an ammeter, and the deflection of the moving coil is dependent upon the product of the current and voltage, *i.e.*, upon the power. On an a.c. circuit it is dependent also upon the phase relationship between current and voltage, so that the power factor is taken automatically into account.

Meters for Measuring Resistance

For the rough measurement of resistance without the use of a Wheatstone bridge, a combination called an **ohm-meter** is sometimes employed. It consists of a moving-coil meter and a source of e.m.f. such as a small cell. Provided that the e.m.f. remains constant, the current flowing depends upon the external resistance to which the combination is applied, and the scale of the meter can therefore be marked in ohms. Only comparatively high resistances can be measured, and owing to the difficulty of compensating for changes in the e.m.f., great accuracy is not to be expected.

For the measurement of high resistances such as insulation resistances, the instrument most commonly employed comprises a small hand-turned d.c. generator for sending a current through the resistance to be measured, together with a special moving-coil meter. The latter has two coils on the same spindle, a current coil in series with the resistance to be measured and a voltage coil connected across the generator. The arrangement is such that the two coils tend to move the pointer in opposite directions, and the voltage coil is thus able to correct the resistance readings given by the current coil in order to compensate for variations in the voltage produced by the generator.

Galvanometers

A simple type of galvanometer was described in Chapter I, and we have referred several times to the use of such instruments for experimental purposes. Galvanometers are employed both for the detection of current and for comparing one current with another, but as their scales are not marked in any particular units, they do not indicate the value of a current directly.

The example described in Chapter I is only moderately sensitive, and the compass needle is liable to be affected by stray magnetic fields. For laboratory use, therefore, more elaborate types are necessary. Most of these operate on the principle of the moving-coil instruments we have already described, but the suspension of the coil itself may be more delicate.

The most sensitive galvanometers are those of the reflecting type. In these the moving part carries a small mirror instead of a pointer. A narrow beam of light is projected on to the mirror, from which it is reflected in the form of a spot of light on to an evenly divided scale. The beam of light between the mirror and the scale acts as a long, weightless pointer, so that a very small angular movement of the mirror causes a large movement of the spot of light along the scale.

If a galvanometer is not heavily damped, and current flows through the coil for a time so short that the current impulse is over before the movement has progressed very far, the first swing of the pointer (or spot of light) is proportional not to the current, but to the quantity of electricity (current × time) which flows during the impulse. An instrument designed for use in this manner is called a **ballistic galvanometer**. It can be used, for example, for comparing the quantities of electricity stored in two capacitors. If both the capacitors are charged to the same voltage, these quantities, as we saw in Chapter XI, are proportional to the capacitances. It follows that if the value of either capacitor is known, that of the other can be calculated from the galvanometer deflections.

Supply Meters

In order that consumers may be called upon to pay for the energy which they take from public supply mains, we need a meter which will take into account both power (watts) and time (hours). Since, however, the voltage is usually constant, the power on d.c. supplies is proportional to the current, and the meter need take into account only current and time. One such meter is the electrolytic type mentioned on page 42.

In most supply meters, however, a miniature electric motor is operated by the current, its speed being proportional to the power taken at any time. By means of gearing, the revolutions made by the motor are counted and displayed on a series of dials marked in the equivalent number of kilowatt-hours. In a.c. meters, the power factor is taken automatically into account.

QUESTIONS

1. Make a sketch showing how to connect a voltmeter and ammeter for measuring the voltage across an electric lamp and the current flowing through it.
2. Why is it desirable for a voltmeter to have a high resistance?

3. Write a short description of a moving-iron meter.

4. What are the advantages of a moving-coil meter?

5. How is it that it is possible to use the same instrument as either a voltmeter or ammeter?

6. Make a list of different types of ammeter and voltmeter, and say which are suitable for use on alternating-current supplies.

7. What is meant by damping? How is it effected in a moving-coil meter?

8. If the resistance of a meter reading up to 100 milliamperes is 2 ohms, what resistance must be connected across it to enable readings to be taken up to 10 amperes?

9. If the resistance of a meter reading up to 10 volts is 5000 ohms, what series resistance is necessary to adapt it for reading up to 40 volts?

10. Make a sketch showing the action of either a hot-wire ammeter or an electrostatic voltmeter.

11. Explain the action of (a) a wattmeter and (b) a meter for measuring insulation resistance.

12. What is a ballistic galvanometer?

CHAPTER XVII

ELECTRICAL COMMUNICATION

THE use of electricity for purposes of communication is so widespread that our book would be incomplete without some reference to the principles involved in simple telephone and telegraph circuits.

Transmission of Speech

Sound consists of vibrations, usually in the air, which may be compared to ripples on the surface of a pond. Instead of being confined to a surface, however, the air vibrations spread out in all directions from their source. Moreover, while the horizontal movement of a ripple across a pond is produced by up-and-down movements of the particles of water, the vibration of the particles of air takes place in the same direction as that in which the sound is transmitted.

The frequency of the vibrations settles the pitch of the sound, a low frequency producing a low note and a high frequency a high note. The frequency of the notes on a piano ranges from thirty to three thousand cycles per second (30 to 3000 Hz). The ear can detect frequencies of from about 20 to 20,000 Hz.

The sound-waves have a wave form which can be represented by a sine curve (page 121) in simple cases, but which is very complex in the case of speech. The problem of telephony is to reproduce the sound-waves at a distant point. This is done by the use of a transmitter or microphone which converts the sound-waves into corresponding variations in the strength of an electric current, and a receiver which responds to the varying current and generates new sound-waves similar to those which affected the transmitter.

Elementary Telephone

The transmitter commonly employed is a special type of variable resistance controlled by the sound-waves. An elementary form is shown in section on the left of

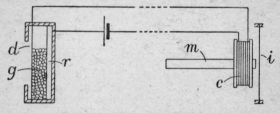

FIG. 113.—Elementary Telephone.

Fig. 113. A flexible conducting diaphragm d is separated from a rigid conducting disc r by a quantity of small granules of carbon g. External connexions are made to the diaphragm and the rigid disc as shown. Current passing through the transmitter has to traverse a large number of more or less imperfect contacts between the carbon granules, and the resistance of this path is very sensitive to variations in pressure on the diaphragm d.

The diaphragm vibrates in response to any sound-waves directed towards it, thus producing a varying pressure on the granules. The resistance of the transmitter is therefore continually altering in accordance with the frequency and wave-form of the sound waves.

The receiver, shown on the right of the diagram,

consists of a magnet m, a coil c, and an iron diaphragm i. The coil is in series with the transmitter and a cell. As the current which the cell is able to send round the circuit is varied by the transmitter, the coil at the receiver produces corresponding variations in the flux of the magnet. The result is to vibrate the iron diaphragm, thus generating new sound-waves similar to those directed towards the transmitter.

It is of interest to note that the receiver itself will act as a transmitter, so that two receivers connected together can be used as a telephone. When sound-waves are directed on to the diaphragm, its movement results in a continual redistribution of the flux produced by the magnet. The changing flux cuts the coil and induces currents in the winding corresponding to the speech-waves. These currents affect the second receiver in the usual manner. Provided that the instruments are fitted with permanent magnets to produce the initial flux, no battery is required.

Practical Transmitters and Receivers

Many modern commercial transmitters of the variable-resistance type employ two diaphragms, an outer one for receiving the sound-waves and a small inner one forming part of the granule chamber. The two are joined at their centres. The carbon granules are situated between two carbon discs, one attached to the back of the inner diaphragm and the other to the back of the granule chamber. All the contact points influenced by the sound-waves are thus between one carbon surface and another.

For special purposes, transmitters based upon other principles are employed. By way of example we may mention moving-coil microphones, in which a coil coupled to the diaphragm moves in a magnetic field, and thus has speech currents induced in it, and electrostatic microphones in which the diaphragm forms one plate of a capacitor the capacitance of which varies in accordance with the speech waves.

Commercial receivers do not differ in principle from the simple form shown in Fig. 113, but the magnetic circuit is improved by the use of a magnet of horseshoe or some equivalent form. Both the pole-pieces can thus be brought close to the diaphragm. A separate coil is usually fitted to each pole-piece, the two coils being connected in series.

Telephone Circuits

A complete circuit between two telephones must make provision for a transmitter and receiver at each end. One possible arrangement is to connect all four instruments in series and to insert a battery at any convenient point in the circuit. If, however, the line is a long one, the resistance of the transmitters will then form only a small part of the whole, and the variations in current caused by the speech waves will be slight.

This difficulty can be overcome by connecting each transmitter in a local circuit in series with its own

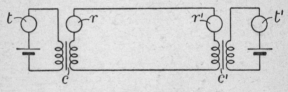

FIG. 114.—Use of Induction Coil.

source of current and the primary winding of an induction coil. The latter is merely a small straight-core transformer of the kind shown diagrammatically in Fig. 63 (page 102). The secondary winding of the induction coil is included in series with the line as shown in Fig. 114, in which the transmitters are shown at $t\ t'$, the receivers at $r\ r'$, and the coils at $c\ c'$.

The variations in the transmitter resistance produce large current variations in the local circuit, and these in turn produce large voltage variations across the secondary winding of the induction coil. The result is a much more effective transmission circuit than would be possible if the transmitters were connected directly in series with the line.

Telephone Exchanges

In a public telephone system each telephone is normally provided with its own pair of wires to the telephone exchange, at which arrangements are made to connect any pair of wires to any other pair. In modern exchange systems local batteries at the telephones are not used, all the transmitters being supplied with current from a common battery at the exchange. Special circuit arrangements are then necessary (a) to retain the advantages of the local transmitter circuit mentioned in the last section and (b) to ensure that the varying speech currents between one pair of telephones connected to the common battery do not reach other telephones drawing their current from the same source. For particulars of these circuits the reader is referred to specialized books on telephony.

Calling is nearly always effected by means of alternating current connected to the line at the exchange. In order that it may respond properly to this current, the bell at the telephone instrument is fitted with a permanent magnet, which causes the armature to move in one direction in response to one half cycle of current and in the opposite direction in response to the next half cycle. Bells of this kind are said to be **polarized.**

Telegraphy

One of the earliest applications of electricity was to the telegraph, and this is still an important means of communication. In its simplest form an electric telegraph consists simply of a source of current, a key for opening and closing the circuit at one end of a line, and a device responsive to the current at the other. Signalling can be carried on over such a circuit by means of the Morse or any other code.

The signal receiving device may be a modified galvanometer, a **sounder** consisting of an electromagnet the armature of which is designed to give a distinctive click at each operation, or a relay arranged to close a local circuit for some other indicating or recording instrument when the line current flows.

Single and Double-Current Working

If the circuit is simply opened and closed at the sending end, it is said to be arranged for **single-current working**. This is satisfactory for short lines on which the capacitance (page 116) is small, but as the capacitance increases, rapid signalling is prevented by the time taken to charge the line when the key contacts are closed and to discharge it when they are opened.

An improvement can be effected by filling up the gaps between one impulse and the next by current flowing in the opposite direction. This is known as **double-current working**, and necessitates the use of a receiving device the response of which is dependent upon the direction in which the current flows. The current constituting the signals proper is called the **marking** current, and that flowing in the opposite direction the **spacing** current. Contacts on the transmitting key are arranged to send a spacing current when the key is at rest and a marking current when it is depressed.

Simplex and Duplex Systems

A **simplex** telegraph circuit is one in which a message can be sent in only one direction at a time. It is possible to send non-interfering messages in both directions at

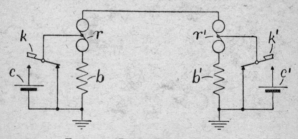

Fig. 115.—Duplex Telegraph Circuit.

the same time by means of what is called a **duplex circuit**. Fig. 115 shows the principle of one form. It is drawn for use with an earth return (page 56).

The signals are received by relays, each of which has

two windings, as represented conventionally at r r'. In addition to its relay, each station has a key k k', a cell c c', and a line balancing resistance b b'. The object of the resistance will appear very shortly.

With the keys in the position shown, no current flows. When key k is pressed, cell c sends a current through the upper windings of both relays, the circuit being completed through the back contact of key k' and the earth return. No current flows in the lower winding of relay r', as this is short-circuited by the key, but the relay is operated by the current in its upper winding.

This current also flows through the upper winding of relay r, but in this case there is another circuit through the lower winding and the line balancing resistance b. The latter is chosen so as to make the current equal to that in the line circuit, and as the two windings oppose each other, the relay does not operate.

Suppose that while key k is still depressed, key k' is depressed also. Cells c and c' are now in opposition, and there is therefore no current in the line. Current flows, however, in the lower windings of both relays, so that relay r operates while relay r' remains in its operated position.

If key k is now released, current from cell c' flows over the line. Relay r is therefore held in its operated position by current in its upper winding, while relay r', which now has current flowing in both windings, releases.

The result is that each relay responds to the movements of the key at the other end of the line and ignores the movements of its own. When either relay operates, its armature closes a circuit (not shown) for any desired form of indicating device.

Direct-current working has now been largely abandoned in telegraph systems, but the duplex circuit is still worthy of notice as an example of the unexpected results obtainable by the application of ingenuity to basic principles.

Voice-Frequency Telegraph Systems

The development of the telephone has greatly influenced the technique of telegraphy, and many modern telegraph systems make use of **voice-frequency currents**, *i.e.*, alternating currents the frequency of which is within the range to which the ear responds. We mentioned in Chapter XIII that combinations of inductance and capacitance possessed a resonant frequency at which a maximum current would flow. By an extension of this principle it is possible to design circuits that will accept currents of one frequency while rejecting those of another. Several messages can therefore be sent over the same line by using a different frequency for each, the various frequencies being sorted out at the receiving end and passed to separate receiving apparatus.

QUESTIONS

1. What is the general principle upon which the telephone is based?
2. Show how speech can be transmitted by means of a simple telephone.
3. Why is it desirable for a microphone to form part of a local circuit instead of being connected directly in series with the line wires?
4. How can a telephone receiver be used as a transmitter?
5. What is meant by single-current and double-current working in telegraphy?
6. What is meant by simplex and duplex systems in telegraphy?

CHAPTER XVIII

THE NATURE OF ELECTRICITY

OUR survey of electrical principles can best be concluded by a short account of the probable nature of electricity itself. Much must of necessity remain unexplained, and that is why we are dealing with the subject at the end instead of at the beginning of our book. Nevertheless, we shall be able to gain some idea of the attitude of modern physicists towards electrical phenomena.

Static Electricity

Although most of our time has been spent in studying

electric currents, we have met the idea of a stationary charge of electricity in examining the action of capacitors. That electricity can exist at rest is shown by the well-known experiment in which a glass rod is rubbed with a silk handkerchief and is then found to have acquired the property of momentarily attracting small pieces of paper. The word "electricity" is derived from the Greek name for amber, which is one of the many other substances exhibiting a similar effect. The electrification of the glass rod was at one time ascribed to a "positive charge" of electricity, a corresponding "negative charge" being produced on the silk, but a more satisfactory explanation is now possible.

The Electron Theory

We referred in Chapter IV to the fact that matter (solids, liquids and gases) is composed of molecules, and that molecules are built up from atoms, of which there are various kinds each corresponding to a different element. In order to explain the nature of electricity, we must push our investigation still further and inquire into the structure of the atom.

Although the atom is far too small to be seen by the aid of the most powerful microscope, it can be studied by indirect methods, and work of this kind has led to the belief that even the atom itself is a collection of smaller particles. These are electrical in nature, and can hardly be regarded as matter at all except when they are associated to form an atom. They are the bricks of which the universe is built, and it is to them that we must look for an explanation of electrical phenomena.

An atom appears to comprise at least two kinds of particle. There is a positive kind called a **proton**, and a negative kind called an **electron**. We need not attach any special significance to the terms positive and negative, except as an indication that the electrical natures of the two kinds of particle are in some way equal and opposite.

In its normal condition an atom has equal numbers of protons and electrons, the number varying according to the element. The simplest atom is that of hydrogen, which has one proton and one electron.

In each atom the protons are collected in a small central nucleus. The relation of the electrons to the nucleus is rather like that of the planets to the sun; this is not a complete analogy, but it will suffice for our present purpose. In the most complicated atom, that of the element uranium, there are ninety-two protons and an equal number of planetary electrons. Each of the elements has its own characteristic number.

In addition to protons, the nucleus may contain uncharged particles known as neutrons. Variations in the number of neutrons give rise to the modified forms of the elements called isotopes, but with these we are not concerned.

The protons being collected in the nucleus, the latter may be looked upon as a positive charge of electricity. This is balanced by the negative charge of the planetary electrons, so that the atom as a whole is neither positive nor negative.

In some substances some of the planetary electrons can be easily detached from the nucleus, and these electrons can move freely between the atoms. Normally this movement takes place in a random manner, so that there is no effective flow of electrons in any particular direction.

Electron Current

A conductor is a substance in which electrons can move readily from atom to atom in this manner; an insulator is one in which they can do so only with difficulty. The effect of applying a potential difference to a conductor is to direct the interchange of free electrons, causing them to be handed on from one atom to the next along the length of the conductor. It is this *directed* movement of electrons which we call an electric current.

A current of 1 ampere is equivalent to a movement of about six million million million electrons per second, and this is therefore the number of electrons representing one coulomb. If the whole population of the earth counted day and night at the rate of one electron each per second, it would take about a hundred years to count them.

The electrons tend to move towards the positive pole of the source of e.m.f. The movement is therefore in the direction opposite to that in which the electric current is conventionally assumed to flow. This is unfortunate, but the current direction was fixed arbitrarily before the true nature of the current was appreciated, and to change the accepted convention now would be very difficult. In the comparatively few cases in which it matters which way the electrons really move, we must remember to make it clear whether we are referring to the actual electron current or to the conventional current which is supposed to flow in the opposite direction.

Insulators

Although in insulators there are few free electrons and it is difficult to detach them from the atoms and thus to cause a current to flow, the application of an e.m.f. does cause a state of strain in which the electrons, while remaining faithful to their parent atoms, are displaced towards the positive pole of the source of e.m.f. This is what happens in the dielectric of a capacitor. If the e.m.f. is increased until a substantial number of electrons are detached, an appreciable current passes, and we say that the insulation has broken down.

Magnetism

The individual atoms of a substance appear to act as tiny magnets; one explanation of this is that the movement of the electrons is equivalent to current flowing in a coil. In the normal condition of an unmagnetized piece of iron, these atomic magnets, or groups of them, are arranged in a haphazard manner, and the iron as a whole does not exhibit any magnetism. The process of magnetization consists in the alignment of the tiny magnets so that all the north poles point in one direction and all the south poles in the other. The iron as a whole then acts as a magnet. The effect which we called saturation in Chapter VII occurs when all the atoms have been brought into alignment.

Ionization

If an atom loses one of its planetary electrons, there are no longer the right number to balance the positive charge of the nucleus. In this condition the atom is called a **positive ion**. If, on the other hand, an atom gains an electron, there are more than enough to balance the positive charge, and the atom is called a **negative ion**. A positive ion will attract free electrons, and a negative ion will readily discard its surplus, the tendency in both cases being to return to the neutral state.

Not only individual atoms, but whole bodies, can exhibit these effects, and the experiment of rubbing the glass rod with the silk handkerchief can be interpreted by saying that electrons are transferred from the glass to the silk. The result is a deficit of electrons, or positive charge, on the glass and a surplus of electrons, or negative charge, on the silk.

Conduction in Electrolytes

We have seen that pure water is a very poor conductor, while a solution is, in general, a comparatively good one. It is supposed that when a substance is dissolved, some of its molecules split into separate ions, one part having an excess of electrons, and the other a deficit.

The case of the copper-sulphate solution mentioned in Chapter IV will serve as an example. A molecule of copper sulphate consists of one atom of copper, one of sulphur and four of oxygen. When the copper sulphate is dissolved, the molecules split up into copper ions and " sulphate " (sulphur plus oxygen) ions. The former have a deficit of electrons and are therefore positive, while the latter have an excess and are therefore negative.

When electrodes are immersed in the solution and an e.m.f. is applied, the positive ions move towards the negative pole (cathode) and the negative ions towards the positive pole (anode). The negative ions give up their surplus electrons at the anode, and the positive ions have their deficit made good at the cathode. There is therefore a passage of current through the electrolyte, while the neutral atoms resulting from the arrival of the ions at the anode and cathode give rise to the chemical effects described in Chapter IV.

Conduction in Gases

In their normal condition, gases are almost perfect insulators. They can, however, be ionized, and an appreciable current can then be made to flow.

A sufficiently high potential difference between electrodes in a glass tube containing gas at a low pressure will produce ionization. One result is to cause the gas to glow, and use is made of this effect in the tubes used in advertising signs. The colour of the light depends upon the kind of gas employed.

The arc lamp provides another example of the conduction of current through a gas. Very high voltages are necessary to break down the insulation between two electrodes separated by air, but if the electrodes are brought together and then separated, a comparatively small voltage will maintain an arc between them. Except in a few special applications, arc lamps have been replaced by other types, but the electric arc is widely used as a heating agent in welding processes, while its destructive effect is a factor to be guarded against in the design of switchgear.

Conduction in Vacuum

Current can also be made to pass through a vacuum, but in this case there is no gas to be ionized and the flow is one of electrons only. On page 152 we mentioned the use as a rectifier of a thermionic valve having one hot and one cold electrode. The action of this is explained by the emission or throwing off of electrons from the hot electrode and their subsequent passage across the intervening space to the cold one. Electron emission from a heated electrode has made possible the valves used in wireless signalling, but a consideration of these is outside the scope of the present volume.

QUESTIONS

1. What do you understand by (a) a molecule, (b) an atom?
2. What do you know of the constitution of the atom?

3. What do you understand by (*a*) an electron, (*b*) a proton?

4. How does the electron theory explain the high resistance of an insulator?

5. What happens when a piece of iron is magnetized?

6. What is meant by ionization?

7. How is it supposed that current passes through (*a*) a solid conductor and (*b*) an electrolyte?

PRINCIPAL ELECTRICAL UNITS

Name	Unit of	Page
Coulomb (C)	Quantity	20
Ampere (A)	Current	20
Ohm (Ω)	Resistance	21
Siemens (S)	Conductance	28
Volt (V)	E.m.f. and p.d.	32
Joule (J)	Energy	68
Watt (W)	Power	68
Weber (Wb)	Magnetic Flux	76
Tesla (T)	Flux Density	76
Henry (H)	Inductance	103
Farad (F)	Capacitance	110
Hertz (Hz)	Frequency	121

Prefixes

Mega- 1,000,000   Kilo- 1,000

Micro- $\dfrac{1}{1,000,000}$ Milli- $\dfrac{1}{1,000}$

ANSWERS TO NUMERICAL QUESTIONS

Page 20.

    (9) 16 milliamperes.   (10) 2·675 amperes.

    (11) 0·5 coulomb per second.

Page 31.

    (7) 9 ohms.   (8) 100 ohms.

    (9) 90 ohms.   (10) 1 ohm, 0·24 ohm.

    (11) 1·25 ohms.   (12) 4 ohms.

    (13) 3 megohms.   (14) 27 ohms.

    (15) 4·167 ohms, 4·2 ohms.

Page 38.

    (5) 0·02 ampere.   (6) 1000 volts.

    (7) 0·33 ampere.   (8) 80 volts.

    (9) 0·1 ohm.   (10) 4 amperes.

    (11) 75 ohms, 150 ohms.

    (12) 0·25 ampere, 2·25 amperes, 135 volts.

Page 61.

    (2) 1 volt.   (3) 1 ampere, 0·625 ampere.

    (8) 0·6 ohm.   (9) 0·3 ohm.

    (10) 4 volts, 4 ohms, 0·2 ampere.

Page 70.

    (3) 3 times.   (11) 3 amperes, 0·72 unit.

    (12) 40 hours.   (13) 960 ohms.

    (14) 0·2 pence.   (15) 0·09 pence.

Page 81.

    (11) 0·4 ampere.   (12) 5 volts.

    (13) 2000 turns.

Page 89.

    (4) 0·5 tesla.

Page 100.

    (10) 9 amperes   (11) 1404 watts

Page 117.

    (7) 0·00026 microfarad.

    (12) 0·167 microfarad, 0·75 microfarad.

    (13) 0·1 microfarad.   (14) 0·25 microfarad.

Page 125.

    (7) 325 volts.   (10) 4·95 amperes.

   (11) 100 times.

Page 134.

    (4) 0·8.   (6) 500 ohms.   (7) 0·6.

Page 145.

    (7) 346 volts.

Page 164.

    (8) 0·0202 ohm.   (9) 15,000 ohms.

INDEX

INDEX

Also in Teach Yourself Books

CALCULUS

P. Abbott

This book has been written as a course in calculus both for those who have to study the subject on their own and for use in the classroom.

Although it is assumed that the reader has an understanding of the fundamentals of algebra, trigonometry and geometry, this course has been carefully designed for the beginner, taking him through a carefully graded series of lessons. Progressing from the elementary stages, the student should find that, on working through the course, he will have a sound knowledge of calculus which he can apply to other fields such as engineering.

A full length course in calculus, revised and updated, incorporating SI units throughout the text.

United Kingdom	60p
Australia (recommended)	$1·80
New Zealand	$1·70
Canada	$2·50

ISBN 0 340 05536 7

ELECTRONICS

W. P. Jolly

This book begins with a review of modern electronics, presenting the fundamental concepts of the electron, energy and waves, and the principles applied in communication, computation and control systems.

Subsequent chapters describe the nature and properties of such materials as the valve, the junction diode, the transistor, and their applications. This leads to the nature and functions of electronic circuits and specific systems like computers. Also examined are developments outside the mainstream of electronics—lasers, cryoelectronics and thyristors.

This introduction to electronics is recommended for students of general science and 'A' level Physics, as an introductory text for students of degree and HNC/HND courses in Physics and Electrical Engineering and for all science students requiring a background knowledge of electronics.

United Kingdom	60p
Australia (recommended)	$1·80
New Zealand	$1·70
Canada	$2·50

ISBN 0 340 19410 3

PHYSICS

David Bryant

This book offers a complete and unified guide to elementary
physics for the interested layman who requires a modern
approach to the subject.

In the form of a reader, PHYSICS includes lucid dis-
cussions of molecular and atomic structure, forces, energy
and waves, and of the behaviour of light, gases and electricity. The text is illustrated with numerous sketches and
photographs, and no previous knowledge is assumed on the
part of the reader beyond a familiarity with basic mathematics.

'An exceptionally readable account of physics up to "O"
level standard.'

The Times Educational Supplement

United Kingdom	50p
Australia (recommended)	$1·50
New Zealand	$1·45
Canada	$0·95

ISBN 0 340 15251 6

RADIO SERVICING

L. Butterworth

This progressive course of instruction covers both the
theoretical and practical aspects of Radio Servicing, and is
designed to provide the beginner with that fundamental
knowledge which he must acquire before venturing into the
wider fields of frequency modulation, high fidelity equipment and television. The book is thus an introduction to all
these subjects as well as to the craft of Radio Servicing.

United Kingdom	50p
Australia (recommended)	$1·50
New Zealand	$1·45
Canada	$1·95

ISBN 0 340 05701 7